Beijia Ning
Analog Electronic Circuits

Information and Computer Engineering

Volume 1

Beijia Ning

Analog Electronic Circuits

—

DE GRUYTER   Science Press
Beijing

Author
Associate Professor Beijia Ning
Xidian University
School of Electronic Engineering
XI'AN, China
bjning@xidian.edu.cn

ISBN 978-3-11-059540-6
e-ISBN (PDF) 978-3-11-059386-0
e-ISBN (EPUB) 978-3-11-059319-8
ISSN 2570-1614

Library of Congress Control Number: 2018941442

Bibliographic information published by the Deutsche Nationalbibliothek
The Deutsche Nationalbibliothek lists this publication in the Deutsche Nationalbibliografie;
detailed bibliographic data are available on the Internet at http://dnb.dnb.de.

Preface

In 2008, the School of International Education (SIE) at Xidian University (XDU) enrolled its first grade of international undergraduate students, majoring in electronics and telecommunication. This was the start of regular international undergraduate education at XDU, and throughout China, more and more universities and colleges started enrolling international students. By 2017, over 489 thousand foreign students had studied in the nation; and this number will definitely be greater in 2018.

Also, in 2008, the author began to teach the course of "Analog circuits" for international students. Now, with 10 years of teaching experience with international students, it is the right time to compile a textbook that is specifically intended for the course. This is the first reason for doing the work.

In May 2014, China Science Publishing & Media Limited and XDU organized a committee to arrange a textbook series of English courses for higher education in the information field under the 13th 5-year plan. The textbook of *Analog electric circuits*, along with another 11 proposals, has been given publishing priority. This is the second reason.

In recent years, as computers, the Internet and smart phones influence almost every aspect of our lives, the procedure of higher education should include the latest progress of IT techniques. Besides the traditional printed version, the newly-edited textbook will include soft copies, electric slides, animations of figures and QR codes for supplementary reading materials. Also, readers can comment on the textbook and share their experience with others through smart phones. Diversity of forms is the dominant feature of the textbook. This is also the author's motivation to compile this book.

Now, at the beginning of the year 2018, China Science Publishing & Media Limited announced the cooperation with De Gruyter, one of the most famous publishers in the world possessing over 40,000 titles from nearly 270 years of academic publishing, to present the book series worldwide. This is indeed great encouragement to the author to complete the work.

Last but not least, the purposes of this textbook are as follows: (1) to cover the fundamentals in the field of analog signals, devices and circuits; (2) to demonstrate the basic applications of analog devices and circuits; (3) to introduce the latest developments of analog devices and their applications; (4) to provide a concise and comprehensive textbook of analog circuits for international students, Chinese students receiving a bilingual education and those interested in analog devices and circuit applications.

Beijia Ning
March 2, 2018

https://doi.org/10.1515/9783110593860-201

Contents

Preface —— V

1   The analog world and semiconductors —— 1
1.1   Introduction —— 1
1.1.1   Definition of analog signals —— 1
1.1.2   The existence of analog signals —— 2
1.1.3   Analog vs. digital —— 3
1.1.4   A brief history of analog devices —— 4
1.2   Semiconductor materials —— 6
1.2.1   Intrinsic semiconductors —— 7
1.2.2   Extrinsic semiconductors —— 8
1.3   Chapter summary —— 10
1.4   Questions —— 11

2   Semiconductor diodes and their applications —— 13
2.1   *p-n* junction —— 13
2.2   Semiconductor diodes —— 14
2.3   The Schottky diode —— 17
2.4   Light-emitting diodes —— 18
2.5   Zener diodes —— 21
2.6   Diode applications —— 22
2.7   Chapter summary —— 38
2.8   Questions —— 38

3   Bipolar junction transistors —— 41
3.1   Fundamentals of transistors —— 42
3.1.1   Structure of transistors —— 42
3.1.2   Transistor operation —— 43
3.2   Transistor configurations —— 44
3.2.1   Common-base configuration —— 45
3.2.2   Common-emitter configuration —— 48
3.2.3   Common-collector configuration —— 51
3.3   BJT DC biasing circuits —— 52
3.3.1   Operating point —— 52
3.3.2   Biasing voltages —— 54
3.3.3   Fixed-bias circuit —— 55
3.3.4   Emitter bias circuit —— 62
3.3.5   Voltage-divider bias circuit —— 66

3.3.6	Common-base configuration —— 71	
3.3.7	Common-collector configuration —— 73	
3.4	BJT AC analysis —— 74	
3.4.1	Introduction to AC analysis —— 74	
3.4.2	AC equivalent circuits —— 76	
3.4.3	Transistor r_e model —— 79	
3.4.4	CE Configuration with fixed bias —— 84	
3.4.5	CE configuration with voltage-divider bias —— 88	
3.4.6	CE configuration with emitter bias —— 91	
3.4.7	Emitter-follower configuration —— 94	
3.4.8	Common-base configuration —— 97	
3.5	Chapter summary —— 100	
3.5.1	BJT summary —— 100	
3.5.2	DC Biasing summary —— 101	
3.5.3	AC Analysis summary —— 102	
3.6	Questions —— 104	

4 **Field-effect transistors —— 107**
4.1	Fundamentals of the junction field-effect transistor —— 109	
4.1.1	Constructions of JFETs —— 109	
4.1.2	Biasing conditions of JFETs —— 111	
4.1.3	JFET transfer characteristics —— 115	
4.2	Fundamentals of depletion-type MOSFETs —— 118	
4.2.1	Construction —— 119	
4.2.2	Operation and characteristics —— 120	
4.2.3	Symbols —— 122	
4.3	Fundamentals of enhancement-type MOSFETs —— 123	
4.3.1	Construction —— 123	
4.3.2	Operation and characteristics —— 124	
4.3.3	Symbols —— 128	
4.4	JFET DC biasing configurations —— 129	
4.4.1	Fixed-bias configuration —— 129	
4.4.2	Self-bias configuration —— 134	
4.4.3	Common-gate configuration —— 138	
4.4.4	Voltage-divider configuration —— 141	
4.5	MOSFET DC biasing configurations —— 147	
4.5.1	Depletion-type MOSFETs —— 147	
4.5.2	Enhancement-type MOSFETs —— 157	
4.6	FET AC analysis —— 164	
4.6.1	Transconductance —— 165	
4.6.2	Input impedance —— 166	

4.6.3   Output impedance —— 167
4.6.4   FET AC equivalent circuit —— 167
4.7   JFET AC analysis —— 168
4.7.1   Fixed-bias configuration —— 168
4.7.2   Self-bias configuration —— 172
4.7.3   Voltage-divider configuration —— 176
4.7.4   Source-follower configuration —— 178
4.7.5   Common-gate configuration —— 182
4.8   MOSFET AC analysis —— 186
4.8.1   Depletion-type MOSFETs —— 186
4.8.2   Enhancement-type MOSFET —— 188
4.9   Chapter summary —— 195
4.9.1   JFET and MOSFET Summary —— 195
4.9.2   DC biasing summary —— 198
4.9.3   AC analysis summary —— 200
4.10   Questions —— 202

5   Operational amplifiers and their applications —— 205
5.1   Overview —— 205
5.2   Fundamentals of op-amp —— 206
5.3   Common-mode rejection ratio —— 211
5.4   Op-amp basic operations —— 216
5.4.1   Transfer characteristic —— 216
5.4.2   Basic operations —— 218
5.4.3   Golden rules of op-amps —— 220
5.4.4   Basic op-amp circuits —— 221
5.5   Linear applications of op-amps —— 225
5.5.1   Voltage summing —— 226
5.5.2   Voltage subtraction —— 227
5.5.3   Integrator —— 229
5.5.4   Differentiator —— 232
5.5.5   Instrumentation amplifier —— 233
5.5.6   Active filters with op-amps —— 235
5.6   Nonlinear applications of op-amps —— 244
5.6.1   Comparators —— 244
5.6.2   Schmitt trigger —— 247
5.6.3   Noninverting Schmitt trigger —— 248
5.6.4   Inverting Schmitt trigger —— 251
5.6.5   Precision rectifier —— 253
5.6.6   Logarithmic amplifier —— 258
5.6.7   Exponential amplifiers —— 260
5.6.8   Oscillators —— 261

5.7   Chapter summary —— **263**
5.8   Questions —— **265**

Bibliography —— **269**

Index —— **271**

1 The analog world and semiconductors

1.1 Introduction

1.1.1 Definition of analog signals

An analog (or analogue) electric signal is defined as a time-continuous electric signal for which the time-varying feature (amplitude) is a representation of some other time-varying quantity. In other words, the amplitude of analog electric signal looks like other time-varying phenomena. For example, the air temperature in some place can be measured with a thermometer for 24 h and recorded as shown in Fig. 1.1 (a). It can be seen that the curve is time continuous and the amplitude ranges between roughly 19 and 30 °C. On the other hand, the air temperature can also be measured by a temperature sensor and recorded as its output voltages in the form of an analog electric signal, as shown in Fig. 1.1 (b). It is obvious that the instantaneous voltage of the signal changes continuously with the temperature of the surrounding air. So, the shape of the voltage curve looks like that of the temperature. Moreover, the voltage can be used to represent the temperature [1].

An analog signal is a measured response to changes in physical phenomena, such as force, energy, torque, pressure, sound, light, temperature, position and motion. Normally, an analog signal is achieved using a transducer or a sensor. Analog electric

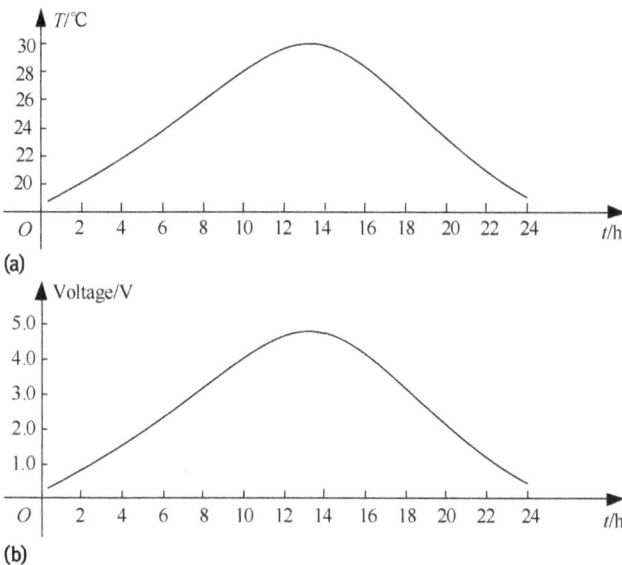

Fig. 1.1: Air temperatures and their measurements, (a) Temperature record during 24 hours, (b) Sensor output voltage during 24 hours.

https://doi.org/10.1515/9783110593860-001

circuits are those that generate, receive and process analog signals with general electric components, such as resistors, inductors and capacitors, or analog devices, such as diodes and transistors. Normally, a practical electric circuit contains both analog and digital parts.

1.1.2 The existence of analog signals

In recent years, there has inevitably been a "digitization" trend, that is, to perform signal processing in digital domains due to the convenience of device use and flexibility of design. The influence of analog circuit analysis and design appears to have diminished. However, this is an illusion. More and more signal processing can be done digitally, but analog circuits will not disappear [2]. The main reason for this is detailed below.

The physical world is an analog place and the analog circuit is the first step for humans to interact with the world. For example, in the real world, as shown in Fig. 1.2, when we speak, the sound wave is a continuous vibration of air. The analog electric signal is first generated by a sensor from the sound wave. Here, the sensor is a microphone. Then, the analog signal is fed into an analog-to-digital converter (A/D) to generate a digital signal. Actually, a digital signal is a sequence of 1 s and 0 s, a numerical representation of the analog signal. It may be easier and more cost efficient to process signals in the digital world. This is called digital signal processing (DSP), a mathematical technique to perform transformations or extract information. Then, if necessary, a digital signal can be brought back out to the analog world through the digital-to-analog converter (D/A). Now, we can hear the voice from a loud-speaker. It is again an analog signal. When we contact the real world, we use analog signals.

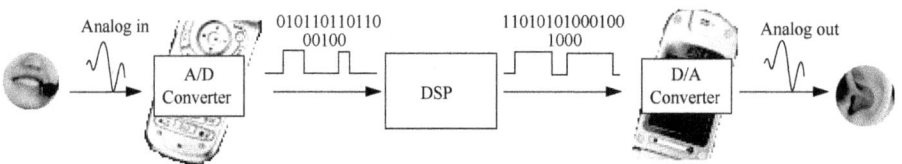

Fig. 1.2: The analog signal and digital signal processing.

Moreover, there are many mature analog building blocks such as operational amplifiers, transistor amplifiers, comparators, A/D and D/A converters, phase-locked loops (PLL) and voltage references (to name just a few) that are still in use and will be used far into the future.

Another reason that makes some users ignore analog circuits is that many latest type DSP integrated circuits (ICs) contain an analog circuit part; in other words, analog circuits are integrated peripherals of a digital IC to serve some interface functions.

More than this, some sensors also contain A/D convertors to output digital signals for convenience.

No matter what kind of designer one wants to be, analog or digital, analog signals and circuits will always exist. Without sufficient knowledge of analog signals and circuits, no one can be a qualified circuit designer.

1.1.3 Analog vs. digital

Analog and digital signals are two different ways to describe the world, the former directly, while the latter indirectly [2]. The differences between them is obvious. For analog signals, as shown in Fig. 1.1 (b), we can see the following:
(1) It may appear at any time instant. The time is dense, from 0 o'clock to 24 o'clock.
(2) The values are real numbers and they are uncountable. Here, from 0 to 5 V.
(3) Small fluctuations in analog signals are meaningful, for example, 3.99 V ≠ 4.00 V.
(4) It is easily affected by noise. A noise contaminated signal is shown in Fig. 1.3. The noise makes the signal fluctuate, away from original values.

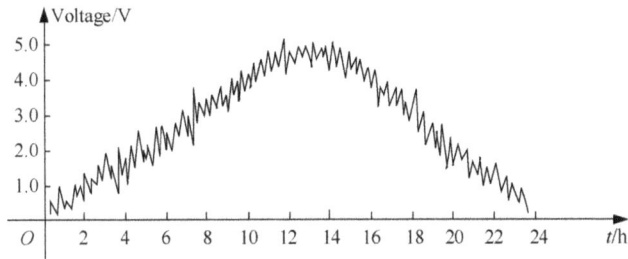

Fig. 1.3: Noise contaminated temperature record.

The digital signal, on the other hand, is a mathematical representation of the analog signal. Here, we assume that the sampling interval of time is 1 h and the amplitude is quantized as unsigned 8-bit binary number. As shown in Fig. 1.4 (a), the 24 sampled original values are 0.8 V, 1.0 V, 1.2 V, 1.5 V, 1.8 V, 2.2 V, 2.8 V, 3.1 V, 3.5 V, 4.0 V, 4.4 V, 4.6 V, 4.8 V, 4.7 V, 4.5 V, 4.1 V, 3.8 V, 3.3 V, 2.8 V, 2.2 V, 1.8 V, 1.2 V, 0.9 V and 0.4 V. Then the unsigned 2-bit hexadecimal digital values (the same as the unsigned 8-bit binary numbers) are 29, 33, 3D, 4D, 5C, 70, 8F, 9E, B3, CC, E0, EB, F5, F0, EA, D1, C2, A8, 8F, 70, 5C, 3D, 2E and 14. As long as the difference between two analog values is less than the quantization level, that is $5/255 = 0.0196$ V, the digital values are same. So, 3.99 V and 4.00 V both correspond to digital values of CC. This explains why small fluctuations of values can be ignored by the digital signal. This is also the reason that the digital signal cannot be easily affected by noise, as shown in Fig. 1.4 (b).

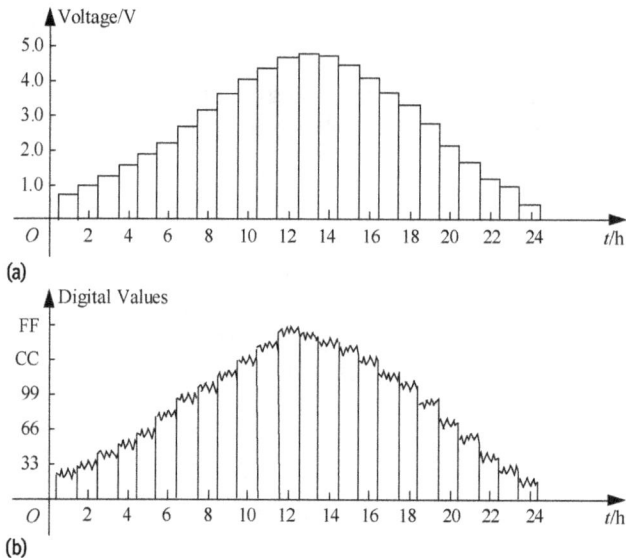

Fig. 1.4: Digitalization of temperature recordings, (a) Digitalized temperature signal, (b) Noise contaminated digital signal.

The properties of digital signal are as follows:
(1) It is meaningful only within sampling intervals.
(2) The values are integers and they are countable.
(3) Small fluctuations in the digital signal are ignored.
(4) It is robust to noise.

To be exact, the digital signal is only an approximate representation of the analog signal. Along the time axis, one can increase the sampling frequency (reduce the sampling interval) to obtain more sample points. On the other hand, along the voltage axis, one can use more bits to represent the voltage to obtain a finer resolution. However, the differences (called quantization errors) exist forever and cannot be totally eliminated. Measures to decrease quantization errors will increase the cost and the complexity of the system. A compromise is needed.

1.1.4 A brief history of analog devices

Electronics deals with the motion of charged particles (electrons and holes) in metals, semiconductors and vacuum. For electronics, what the charged particles transfer is information, instead of energy. It works within a wide range of frequency spectra (several Hertz to Gigahertz) and at relatively low voltages (a few to tens of volts) and weak currents (several mA to a few A) [3–5].

It began in 1895 with the existence of discrete charges from the theory proposed by Lorentz. In 1897, Thomson successfully measured the mass of cathode rays, showing that they consisted of negatively charged particles smaller than atoms, the first "subatomic particles", which were later named electrons [6].

The cathode ray tube (CRT) was invented by the German physicist Braun in 1897. It was a cold-cathode diode with a phosphor-coated screen.

In 1904 the English physicist Fleming invented the vacuum tube, later known as the Fleming valve, which was a basic component for electronics. It could be used as a rectifier of alternating current (AC) and as a radio wave detector, which led to a wide range of applications in radar, television and radio broadcasting, sound amplification, recording and reproduction, telephone systems, analog computers and industrial control [7].

The valve greatly improved the crystal set that rectified the radio signal using an early solid-state diode based on a crystal, the so-called cat's whisker, which was the first type of semiconductor diode, and in fact, one of the first semiconductor electronic devices, developed in around 1904 by Dunwoody, Pickard and others. Unlike modern semiconductors, such a diode required painstaking adjustment of the contact to the crystal in order to rectify. Crystal radios were the most popular type of radio until the mid-1920s when they were replaced by vacuum tube radios. The vacuum tube was a reliable device for detecting radio signals. It was the invention of the vacuum tube that made electronics practical and widespread throughout the first half of the twentieth century.

In the 1940s, the invention of semiconductor devices exploiting the electronic properties of semiconductor materials, principally silicon, germanium and gallium arsenide, made it possible to produce solid-state devices. Compared with vacuum tubes, solid-state devices are smaller, lighter, cheaper, more efficient, more reliable, more durable and more convenient to use without warming up.

The transistor, based on semiconductor technology, practically implemented in 1947 by American physicists Bardeen, Brattain and Shockley, is the fundamental building block of modern electronic devices, and is ubiquitous in modern electronic systems. The transistor is used to amplify or switch electronic signals and electrical power. It revolutionized electronics and paved the way for smaller and cheaper radios, calculators, computers and so on. The transistor is on the list of IEEE milestones in electronics, and Bardeen, Brattain and Shockley shared the 1956 Nobel Prize in Physics for their achievement.

Today, although over a billion individually packaged (known as discrete) transistors are produced every year, the vast majority of transistors are now implemented in integrated circuits (ICs), along with resistors, capacitors, diodes and other electronic components, to produce more complex and multifunctional electronic circuits. For example, a logic gate consists of up to about 20 transistors, whereas an advanced microprocessor, as of the year 2009, can use as many as 3 billion transistors (MOSFETs, to be introduced in Chapter 4).

However, there are still a few applications for which vacuum tubes are preferred to semiconductors, such as high-power RF amplifiers, cathode ray tubes, audio equipment and some microwave devices.

Moreover, based on logic gates, a new branch of electronics has been established, i.e., digital circuits and digital signal processing. This is another story and beyond this textbook.

1.2 Semiconductor materials

By resistivity, materials may be classified as conductors, insulators and semiconductors. The classification depends on the value of the resistivity of the material, which is the resistance of a unit cube of the material measured between opposite faces of the cube. The unit of resistivity is ohm meter (Ωm). Normally, resistivity varies with temperature. Good conductors are usually metals with resistivities in the order of 10^{-7} to 10^{-8} Ωm. Semiconductors have resistivities on the order of 10^{-3} to 3×10^{3} Ωm. The resistivities of insulators are on the order of 10^{4} to 10^{14} Ωm. Some typical material resistivities at normal room temperatures are shown in Tab. 1.1 [3, 8–10].

In general, over a limited range of temperatures, when the temperature increases, the resistance of a conductor increases, that of an insulator remains approximately constant, whereas that of a semiconductor decreases.

Tab. 1.1: Typical approximate resistivity values of materials

	Material	Resistivity (Ωm) at 20 °C
Conductors	Silver	1.59×10^{-8}
	Copper	1.68×10^{-8}
	Aluminum	2.82×10^{-8}
	Carbon (graphene)	1.00×10^{-8}
Semiconductors	Carbon (graphite)	2.50×10^{-6} to 5.00×10^{-6}
	GaAs	1.00×10^{-3} to 1.00×10^{8}
	Germanium	4.60×10^{-1}
	Silicon	6.40×10^{2}
Insulators	Glass	1.00×10^{11} to 1.00×10^{15}
	Carbon (diamond)	1.00×10^{12}
	Hard rubber	1.00×10^{13}
	Air	1.30×10^{16} to 3.00×10^{16}

1.2.1 Intrinsic semiconductors

Why do these materials, when they are in the solid state, have quite different conductivities? Actually, whether the solid is a metal or a dielectric depends not so much on the properties of the atoms forming the crystal, but on the types of bonds of atoms in the crystal lattice of the solid body. For example, the atoms of carbon (C), depending on the different types of crystal they form, can be either good conductors (graphene and graphite), or a perfect insulator (diamond).

Instead of neutral atoms, the metallic crystal lattice is formed by positively charged ions. While forming the lattice, each atom loses one valence electron. As shown in Fig. 1.5 (a) as black dots, these electrons do not belong to any specific ion of the metal. These electrons are said to be collectivized by the crystal and can move freely under the action of the external electric field. These are called free electrons and are carriers of electric currents. There are enormous numbers of electrons in the unit volume of the metal. It is no wonder that metals are perfect conductors. The scheme of the crystal lattice of pure silicon demonstrates another type of bond, as shown in Fig. 1.5 (b). The four valence electrons of one atom form a bonding arrangement with four adjoining atoms. This bonding of atoms, strengthened by the sharing of electrons, is called covalent bonding. Because of this, there is no free particle to move, and as a whole, the material is still a poor conductor.

Being a compound material, gallium arsenide (GaAs) has bonds between two different atoms, as shown in Fig. 1.5 (c). Each atom is surrounded by atoms of the complementary type. The bonds still exist, but now five electrons are provided by the Arsenic atom and three by the Gallium atom. Similarly to pure silicon crystal, GaAs does not have free particles to move either and is still a poor conductor [7].

Semiconductor materials, in the form of single crystal, such as silicon and germanium, or as a compound, like gallium arsenide (GaAs), have stable atomic structures or lattices with a relatively low level of free particles. They are called intrinsic materi-

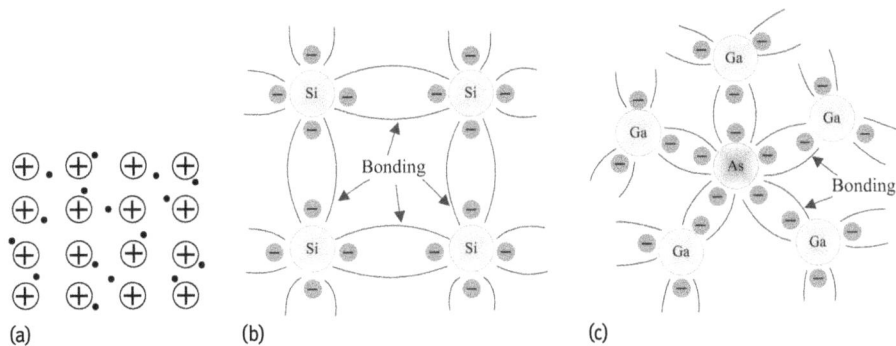

Fig. 1.5: Lattice of three types of materials, (a) Metallic crystal lattice, (b) Silicon (Si) crystal lattice, (c) Lattice of GaAs.

als if they are refined to reduce the number of impurities to a very low level, as low as current technology can. Intrinsic materials are poor conductors.

1.2.2 Extrinsic semiconductors

As described earlier, intrinsic materials are poor conductors, but their characteristics can be modified greatly by the addition of specific impurity atoms to the relatively pure semiconductor material. Adding extremely small amounts of impurities to pure semiconductors in a controlled manner is called doping. Although only added at 1 part in 10 million, these impurities can alter the atomic lattice sufficiently to change the electrical properties of the material completely [3].

A semiconductor material that has been subjected to the doping process is called an extrinsic material. There are two types of extrinsic materials, n-type and p-type materials, which are of great importance to semiconductor industry. Both types of materials are created by adding a predetermined number of impurity atoms to a silicon base [5].

1. n-type material

An n-type material is formed by introducing impurity elements that have five valence electrons, such as antimony (Sb), arsenic (As) and phosphorus (P), with one more electron than silicon atoms. For example, Fig. 1.6 illustrates the effect of impurity ele-

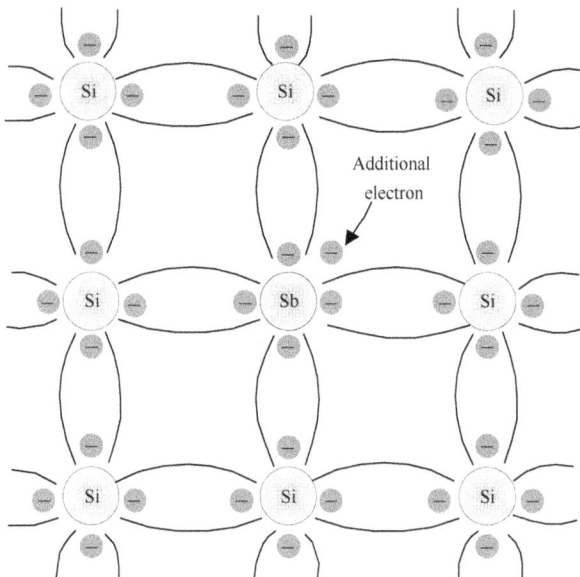

Fig. 1.6: Forming n-type material by adding antimony (Sb) in a silicon (Si) base.

ments. It can be seen that the four covalent bonds still exist between silicon atoms, the same as those in pure silicon crystals. However, there is an additional fifth electron that comes from the impurity antimony atom, which is not linked with any particular covalent bond. This fifth electron is loosely bound to its parent (antimony) atom, which leads to relatively free movement as a negative particle in the material, called *n*-type material [7].

The inserted impurity atom donating a relatively "free" electron to the structure is called a donor atom. Note that even though a large number of free electrons have been brought into the *n*-type material, it is still neutral electrically, since in the nuclei the number of protons is still equal to that of electrons in the structure.

2. *p*-type material

Opposite to *n*-type material, *p*-type material is created by doping a pure germanium or silicon crystal with impurity atoms having three valence electrons, one fewer electron than that of the base material. Boron, gallium and indium are the elements most frequently used for this purpose. For example, the effect of the boron atom on a base of silicon is illustrated in Fig. 1.7.

It can be seen that there is one fewer electron from the boron atom to complete the covalent bonds with the silicon atom in the lattice. The resulting vacancy is called a hole and is indicated by a small circle or a plus sign, representing the absence of a

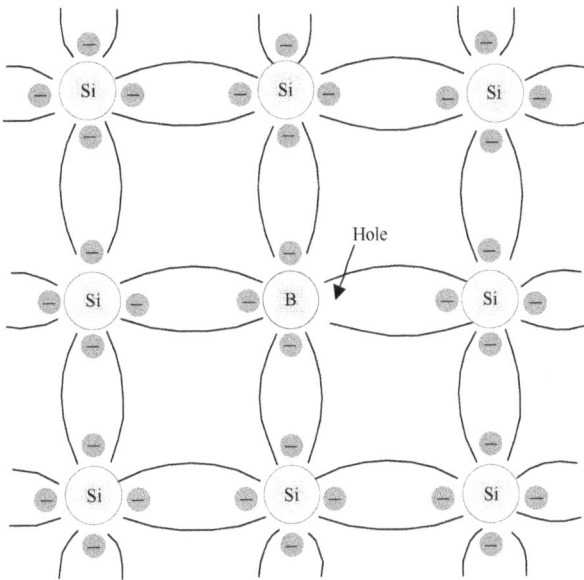

Fig. 1.7: Forming *p*-type material by adding boron (B) in a silicon (Si) base.

negative charge. Since the resulting vacancy will readily accept a free electron from other atoms, the added impurities with one fewer electron are called acceptor atoms.

For the same reasons as described for n-type material, the newly formed p-type material is electrically neutral.

3. Majority and minority carriers

In an n-type material, the number of holes has remained at the same level as in the intrinsic level. In other words, there exist only a limited numbers of holes. Therefore, the net result is that the number of electrons far outweighs the number of holes. For this reason, in an n-type material, as shown in Fig. 1.8 (a), the electron is called the majority carrier and the hole is called the minority carrier.

For the p-type material the number of holes far outweighs the number of electrons, as indicated in Fig. 1.8 (b). So, in a p-type material the hole is the majority carrier and the electron is the minority carrier.

The n- and p-type materials are the basic building blocks of semiconductor devices, which will result in semiconductor devices with considerable importance in the electronic world.

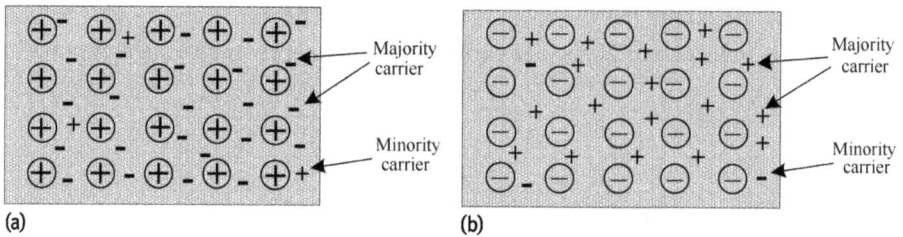

Fig. 1.8: Majority and minority carriers in different types of materials, (a) n-type material, (b) p-type material.

1.3 Chapter summary

In this chapter, as an introduction to analog circuits, the definition of the analog signal is first presented. Then, the generation of analog signal from sensor is discussed, showing its close relationship with real-world applications. Also, as digital technology has been drawing more attention in recent years, analog and digital technologies are compared in several aspects, indicating their advantages, disadvantages and the relationship between them. Moreover, the development history of analog devices is briefly presented, showing that the current analog devices are the results of development throughout all these years and that they will continue to be developed in the future.

Another topic in this chapter is semiconductor materials, which is the base of modern analog devices. From the viewpoint of the conductivity of material, the conductor, semiconductor and insulator are introduced. Then, by the doping process, intrinsic materials are changed to extrinsic materials, leading to different levels of conductivity, which are the building blocks of modern solid-state devices.

1.4 Questions

Q1.1: What are the definitions of analog and digital signals? Which one is closer to the real world? Why and how?

Q1.2: List the main features of analog and digital signals. Pay attention to their differences.

Q1.3: From the Internet and other resources, find the latest development of analog devices in recent years.

Q1.4: Are insulators always insulators? Can they be changed? Give some examples and the conditions for change. Answer the same questions for conductors.

Q1.5: Where do semiconductors come from? List the ways that humans obtain semiconductors.

Q1.6: Give some examples of newly invented materials and their (potential) applications.

Q1.7: How are n- and p-type materials created?

Q1.8: Can any impurity be used in the doping process to create extrinsic material? What conditions should be satisfied?

Q1.9: What are majority and minority carriers? Are they always unchanged in different types of semiconductors?

Q1.10: In recent decades, the development of digital techniques has been very fast and many fields have changed to digital ones. However, some fields are still analog. Give some examples of these fields.

2 Semiconductor diodes and their applications

Now that both *n*- and *p*-type materials are ready here, we are only one step away from the first solid-state electronic device, the semiconductor diode, with its simplicity and wide range of applications, showing its importance for the development of modern solid-state devices.

2.1 *p-n* junction

By simply joining an *n*-type and a *p*-type material together, nothing more, just the joining of one material with a majority carrier of electrons to one with a majority carrier of holes, a *p-n* junction is created with its special conductivity. Simply speaking, a *p-n* junction is a piece of semiconductor material in which part of the material is *p*-type and part is *n*-type [1].

As shown in Fig. 2.1 (a), at the junction, the donated electrons in the *n*-type material, also called majority carriers, diffuse into the *p*-type material, as it is known that diffusion often occurs from an area of high density to an area of lower density. The situation is the same in the *p*-type material: the majority carriers, holes, move across the junction into the opposite side. So, the holes and electrons are combined in the region near the junction, resulting in a lack of free carriers. Once the free carriers have become neutral, only positive and negative ions remain, as shown in Fig. 2.1 (b). So, an electric field is set up from positive ions to negative ions, which prevents the movement of majority carriers to the opposite sides [9].

When balance is reached, a region of uncovered positive and negative ions is generated. This region is called the depletion region, for the reason of "depletion" of free carriers in the region. Across the junction, potential difference exists, which is called the contact potential [10–12].

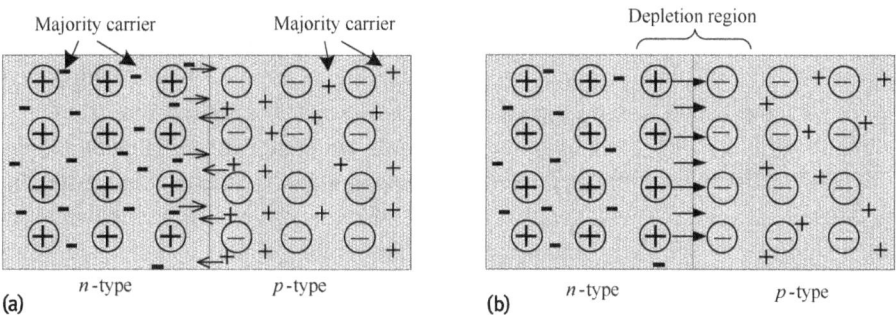

Fig. 2.1: Joining of *n*-type and *p*-type materials, (a) Structure of *p-n* junction, (b) Generation of depletion region.

https://doi.org/10.1515/9783110593860-002

2.2 Semiconductor diodes

A semiconductor diode is a device containing a *p-n* junction mounted in a package with two connecting leads, suitable for conducting and dissipating the heat generated in operation. The structure is shown in Fig. 2.2(a), i.e., an *n*-type material is joined together with a *p*-type material [3].

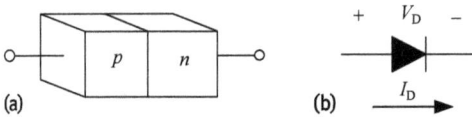

Fig. 2.2: Structure and electronic symbols of a semiconductor diode, (a) Structure of semiconducture diode, (b) Electronic symbol of semiconductor diode.

The electronic symbol for a semiconductor diode is provided in Fig. 2.2(b), showing its correspondence with the *p-n* junction. Note that the polarity of the voltage across the diode and the current direction through it are symbolic. If the voltage applied across the diode is the same as that in Fig. 2.2(b), it is considered a positive voltage. In the reverse situation, it is a negative voltage. The same standards can be applied to the defined direction of current in Fig. 2.2(b) [9–11, 13].

1. No bias condition (V_D = 0 V)
Without any externally applied voltage, the diffusion of majorities will diminish until it stops due to the electric field inside the *p-n* junction. In other words, the diffusion of majorities and the prevention effect of *p-n* junction reaches a balance state.

So, in the absence of an applied bias across a semiconductor diode, the net flow of charge in one direction is zero, i.e., the current under no-bias conditions is zero.

2. Reverse-bias condition (V_D < 0 V)
If an external potential is applied across the *p-n* junction in the way that the positive terminal is connected to the *n*-type material, the negative terminal is connected to the *p*-type material as shown in Fig. 2.3(a). The number of positive ions in the depletion region of the *n*-type material will increase due to the large number of majorities (electrons) drawn to the positive terminal of the applied potential. Similarly, the number of negative ions will increase in the *p*-type material. As a result, the depletion region is widened. This will establish a greater barrier for the majority carriers to overcome, reducing the majority carrier flow to zero. The electrical symbols are shown in Fig. 2.3(b).

On the other hand, the number of minority carriers coming into the depletion region will not change, leading to minority-carrier flow. This current that exists under

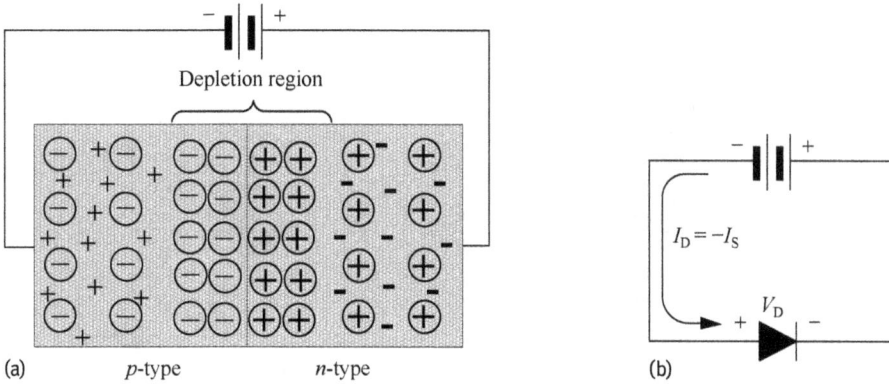

Fig. 2.3: Reverse-bias condition of a semiconductor diode, (a) Internal carrier distribution under reverse-bias condition, (b) Electrical symbols of reverse-bias condition.

reverse bias conditions is called the reverse saturation current and is represented by I_S. The term "saturation" is from the fact that the current reaches its peak value quickly and does not change greatly with increases in the reverse bias voltage. Particularly, note that the direction of I_S is against the symbolic direction of I_D.

Also, the reverse saturation current is normally at the level of a few microamperes, except for high-power devices. Moreover, in recent years, for silicon devices this level has typically been in the nanoampere range.

3. Forward-bias condition ($V_D > 0\,V$)

Applying the positive potential to the p-type material and the negative potential to the n-type material, a forward-bias condition is set up. This is also sometimes called the "on" state, as shown in Fig. 2.4 (a).

The external forward-bias voltage V_D will "force" electrons in the n-type material and holes in the p-type material to recombine with the ions near the junction and thus reduce the width of the depletion region. The reduction in the width of the depletion region will result in a strong majority flow across the junction. The resulting minority-carrier flow of electrons from the p-type material to the n-type material has not changed in magnitude, due to the limited number of impurities in the materials. So, the current in the forward-bias condition is noticeable, as shown in Fig. 2.4 (b).

Now, we briefly summarize the three biasing conditions of the semiconductor diode as shown in Fig. 2.5.

1. When in the no-bias condition, both V_D and I_D are zero.
2. When it is forward biased, the increase of V_D in magnitude will give rise to an exponential rise in current, as shown in Fig. 2.5. Note that the vertical scale is measured in milliamperes for $I_D > 0\,mA$, and the maximum of horizontal scale is 1 V. Therefore, the voltage across a forward-biased diode will typically be less

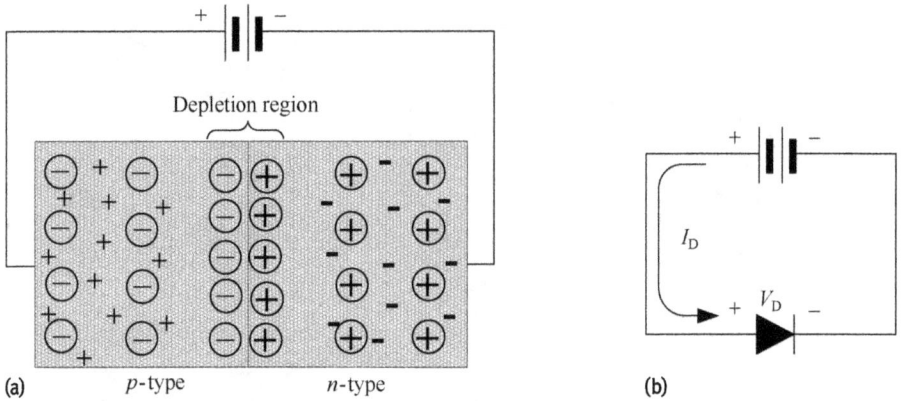

Fig. 2.4: Forward-bias condition of s semiconductor diode, (a) Internal carrier distribution under forward-bias condition, (b) Electrical symbols of forward-bias condition.

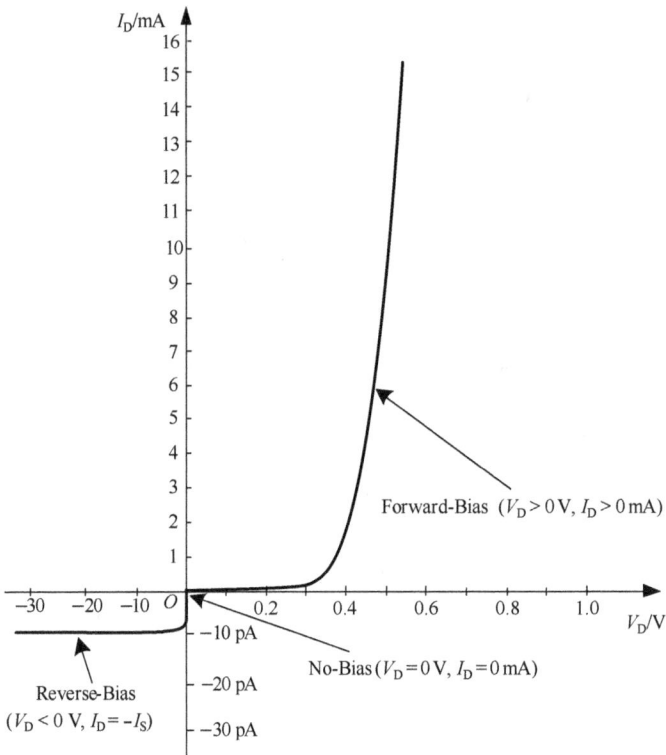

Fig. 2.5: Silicon semiconductor diode characteristics.

than 1 V. It can also be seen how quickly the current rises when V_D is beyond the voltage drop across the diodes. The defined direction of conventional current for the forward voltage region matches the arrowhead in the diode symbol.
3. When in the reverse-biased condition, the direction of I_S is against the symbolic direction of I_D. Also, the reverse current rises quickly and soon reaches the saturation level. However, for a commonly-used diode, the actual reverse saturation current will normally be measurable as larger than that from the curve.

For the forward- and reverse-bias regions, the general characteristics of a semiconductor diode can be defined by the following equation, which is referred to as Shockley's equation:

$$I_D = I_S(e^{V_D/nV_T} - 1) \tag{2.1}$$

where I_S is the reverse saturation current; V_D is the applied forward-bias voltage across the diode; n is an ideality factor, which is a function of the operating conditions and physical construction, ranging between 1 and 2 depending on a wide variety of factors ($n = 1$ will be assumed throughout this textbook unless otherwise indicated). The voltage V_T is called the thermal voltage and is determined by

$$V_T = \frac{kT}{q}$$

where k is Boltzmann's constant = 1.38×10^{-23} J/K; T is the absolute temperature in Kelvin = 273 + the temperature in °C; q is the magnitude of electronic charge = 1.6×10^{-19} C.

2.3 The Schottky diode

Instead of generating a semiconductor–semiconductor junction as in conventional diodes, a metal–semiconductor junction can also be formed between a metal and a semiconductor. Thus, a potential energy barrier for electrons is generated, called the Schottky barrier after Walter Schottky. Schottky barriers have rectifying characteristics, suitable for use as a diode. The electronic symbol of the Schottky diode is shown in Fig. 2.6 [14].

Typically, the semiconductor would be n-type silicon and the metals involved are platinum, molybdenum, tungsten or chromium and certain types of silicides, such as palladium silicide and platinum silicide. The metal side acts as the positive terminal (anode) and the n-type semiconductor as the negative terminal (cathode) of the diode [15].

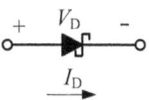

Fig. 2.6: Electronic symbol of Schottky diode.

When applied with forward voltage across the Schottky diode, a current flows in the forward direction. Compared with 600 to 700 mV of forward voltage for a common silicon diode, the Schottky diode has only 150 to 450 mV of forward voltage. This lower forward voltage requirement of the Schottky diode allows higher switching speeds and better system efficiency. So, the applications of the Schottky diode fall into three categories. (1) General purpose: small signal Schottky diodes can be used in many applications as conventional diodes, where a low forward voltage drop or a high switching speed are essential. However, Schottky diodes are more expensive than their conventional counterparts, and this has limited their widespread application. (2) RF mixers: the Schottky diode is the most suitable component for a radio frequency (RF) mixer circuit, in which a deliberate nonlinearity is introduced in order to extract the sum or difference of two input frequencies. The high speed, low noise and large signal handling ability of the Schottky diode make it particularly ideal for wideband mixers. The earliest applications of the Schottky diode were in this field with a range of characterizations for such uses. (3) Rectifiers: the largest application field for the Schottky diode is power supply rectifiers. There is a growing market for medium-to-high current 3.3 V or 5 V-output switchers to supply digital circuits. Greater efficiencies and faster switching speeds of the Schottky diode are the main reasons for this trend [16].

2.4 Light-emitting diodes

A light-emitting diode (LED) is a two-pin p-n junction semiconductor diode, which emits light when applied with external voltage. The electronic symbol of a light-emitting diode is shown in Fig. 2.7. LEDs are normally used as the light source [3].

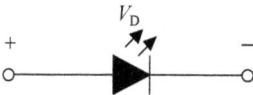

Fig. 2.7: Electronic symbols of a light-emitting diode.

Generally, LED is based on the phenomenon of electroluminescence, which means the emission of light from a p-n junction under an externally applied electric field. In a forward-biased p-n junction, electrons from the n-type region recombine with the holes in the p-type region. The electrons are in the conduction band of energy levels, while holes are in the valence energy band. Thus, the energy level of the holes is less than that of the electrons. During the recombination of electrons and holes, some part of the energy must be dissipated. This energy is sent out in the form of heat and light [12].

Unlike silicon or germanium diodes dissipating energy in the form of heat, gallium arsenide phosphide (GaAsP) and gallium phosphide (GaP) semiconductors dissipate energy by emitting photons. If translucent, the p-n junction becomes the source

Fig. 2.8: LED used as seven-segment display.

of light with emission of light, thus becoming a light-emitting diode. However, when reverse biased, the *p-n* junction will produce no light and, on the contrary, it may even be damaged.

Commercial LEDs were commonly used to replace incandescent and neon indicator lamps in electronics equipment used in laboratories, in the form of seven-segment displays. Later, they were extended to such appliances as radios, TVs, calculators, telephones, as well as watches. A type of widely-used 7-segment LED display is illustrated in Fig. 2.8 [17, 18].

Conventional LEDs are made from a variety of inorganic semiconductor materials, emitting wide range of colors, including infrared, red, orange, yellow, green, blue, violet, purple, ultraviolet, pink and white. White LEDs can now generate over 300 lumens per watt of input electricity with normal lives of up to 100,000 h. Compared to conventional incandescent bulbs, this is not only an enormous increase in power efficiency, but the same or lower cost per bulb for a long run. Fig. 2.9 shows LEDs with different colors used in circuit for information indication.

When the emissive layer of a diode is made from an organic compound of electro-luminescent material, an organic light-emitting diode (OLED) is obtained. The advantages of OLEDs include thinness, low cost, low driving voltage, a broad viewing angle, high visual contrast and a wide color gamut. In recent years, OLEDs have been used to make visual displays for mobile electronic devices such as cellular phones, computer screens, digital cameras and many wearable devices [19].

Fig. 2.9: LEDs used for information indication.

From the very beginning until recent years, LEDs have experienced great development, and their applications fall into four brief categories. (1) Illumination: where light is radiated from source and reflected from objects, to give human visual response, such as LED street lights. (2) Visual signals: on condition that light travels directly from the source to human eyes in order to convey a message or some kind of meaning, such as LED traffic lights. (3) Data communication: light can be used to transmit data and analog signals, not for human eyes. (4) Light sensors: where LEDs function in a reverse-biased mode and respond to incident light, instead of sending out light [17].

Compared with conventional counterparts, LEDs possess many advantages as follows [17]. (1) Efficiency: LEDs can transfer electric energy to light with higher efficiency than incandescent light lamps. Moreover, the efficiency is almost unaffected by their shapes or sizes. (2) Color: LEDs can inherently send out light of a desirable color without using any color filters which traditional lighting methods need, leading to higher efficiency and lower cost. (3) Size: LEDs can be very small for some types of packages, resulting in easy mounting on printed circuit boards. (4) There is no need for a warm-up time: LEDs light up at a very fast speed with a transition period of several microseconds. (5) Flicker: unlike incandescent and fluorescent bulbs that fail quickly when flickered often, LEDs are more durable even with a high frequency of flickering. (6) Brightness control: the brightness of LEDs can be easily controlled by either the forward current or voltage of pulse-width modulation. (7) Cold source: compared with most other light sources, during operation LEDs produce very little heat that may damage surrounding objects or fabrics. Useless heat is dispersed through the LED base. (8) Slow failure: LEDs normally fail by dimming over time, rather than a sudden failure like incandescent bulbs. This gives the user sufficient time to replace old LEDs with new ones in daily maintenance. (9) Lifetime: typically, incandescent light bulbs have lives of about 1,000 to 2,000 h, while fluorescent tubes have a lifetime of about 10,000 to 15,000 h. LEDs can have 35,000 to 50,000 h of useful lives. This leads to low-cost maintenance during the life cycle, which is the dominant benefit that makes LEDs competitive on the market, along with energy savings. (10) Shock resistance: unlike fragile incandescent and fluorescent bulbs, LEDs are solid-state components and are not easily damaged by external shocks. (11) Focus: fluorescent and incandescent sources often need an external reflector to concentrate light, while LEDs can easily focus their light with a specially-designed solid package [3, 12].

Light-emitting diodes are now used in applications as broad as traffic signals, automotive headlamps, aviation lighting, camera/handphone flashes, advertising, light-up wallpaper and indoor lighting. Moreover, in the near future, this trend will not show any signs of slowing down.

2.5 Zener diodes

Going back to Fig. 2.5, we can see that the reverse-biased voltage can be tens of volts. Moreover, there is a point where the application of too negative a voltage will give rise to a sharp change in the characteristics, as the solid line shows in Fig. 2.10. This part of the characteristics is called the Zener region, where the current increases rapidly in a reversed direction. The reverse-bias voltage that leads to this dramatic increase in characteristics is called the Zener potential and is designated as V_Z. Diodes employing this unique portion of the characteristic of a *p-n* junction are called Zener diodes. The Zener region of the general purposes semiconductor diode must be avoided [3].

The position of the Zener region can be manipulated by changing the doping levels. An increase in doping that generates more impurities will result in a decrease of V_Z. Zener diodes are available with V_Z of 1.8 to 200 V. Normally, silicon is the preferred material in the manufacture of Zener diodes, due to its excellent temperature and current capabilities [11].

The unique characteristics in the Zener region are employed in the implementation of Zener diodes, which have the electronic symbol shown in Fig. 2.11 (a) [12].

The direction of conduction of the Zener diode, as I_Z shown in Fig. 2.11 (a), is opposite to that of the arrow in the symbol.

Although there is a slight slope in the characteristics appearing in Fig. 2.10, it would be acceptable to assume the Zener diode as a straight vertical line at V_Z. Therefore, for most of the applications in this textbook, the series resistive element can be ignored and the reduced equivalent model of just a DC battery of V_Z volts can be employed, as shown in Fig. 2.11 (b).

Moreover, in the reverse-bias region below V_Z, the equivalent model for a Zener diode is a very large resistor, the same as for the general-purpose diode. For most applications, this resistance is so large that can be ignored, and the open-circuit can be

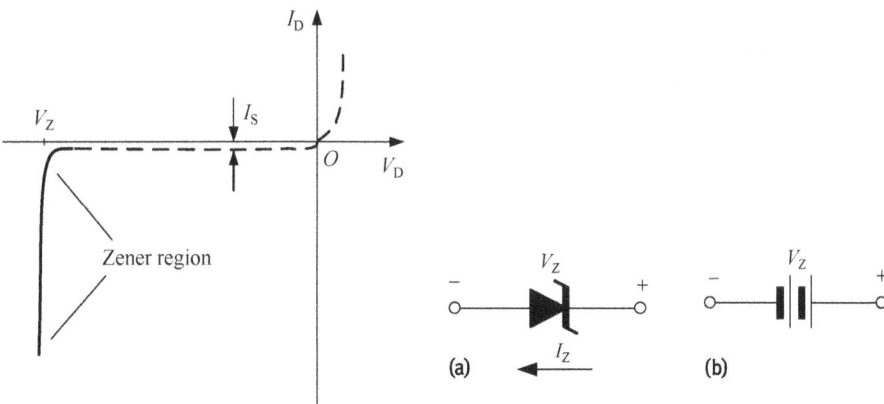

Fig. 2.10: Zener region.

Fig. 2.11: Zener diode, (a) Electronic symbol; (b) Equivalent model.

employed as an equivalent circuit. Also, in the forward-bias region, the properties of the Zener diode are the same as those of the general-purpose diode [16].

2.6 Diode applications

The structure, characteristics and equivalent circuits of semiconductor diodes were introduced in the previous sections. The main goal of this section is to develop a working knowledge of the diode in different configurations using appropriate equivalent circuits for a variety of applications. The concepts introduced in this section will still be significant for the discussion in the following chapters. For example, the construction of diode can also be employed in the description of the basic structure of other types of semiconductor devices, such as bipolar junction transistor sand field effect transistors.

Now that we have a basic knowledge of the characteristics of a diode along with its response to applied external voltages and currents, the examination of a wide variety of circuits using this knowledge can be carried out. In general, the analysis of electronic networks can be done in either of two ways: using the actual characteristics or applying an equipment circuit of the device.

The initial discussion of diodes will include the actual characteristics to demonstrate the characteristics. Then, the approximate piecewise models, or equipment circuits, will be employed without lengthy mathematical derivations. Although the results obtained from the actual characteristics and equivalent circuits may be slightly different, we should not forget that the characteristics obtained from a specification sheet may be slightly different from those of the actual device. Normally, the difference is trivial and sufficient to justify the approximations employed in the analysis. Also, other elements of the network, such as resistor and capacitors, still have some level of tolerances. A response determined through an appropriate set of approximations can always be regarded as accurate as the one that employs the full characteristics [3].

In the discussion in this textbook, the emphasis is on the development of a working knowledge of a device by means of appropriate approximations, so as to avoid unnecessary mathematical complexity. However, occasionally a detailed mathematical analysis will be provided through characteristics.

Example 2.1
Now solve the simplest diode configuration as illustrated in Fig. 2.12 (a), i.e., find the current I_D and voltage V_D that will satisfy both the characteristics of the diode and the parameters of other components in the network at the same time. The actual characteristics of the diode, as shown in Fig. 2.12 (b), will be used to analyze this network [11, 13].

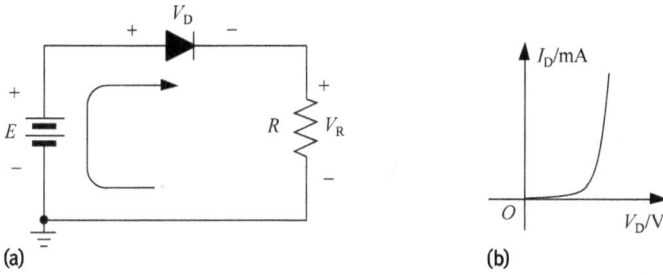

Fig. 2.12: Configuration of a series diode, (a) Circuit, (b) Characteristics.

Solution 1 (graphical method)

First by applying Kirchhoff's voltage law in the clockwise direction in Fig. 2.12 (a), we get

$$E - V_D - V_R = 0$$

Using the relationship between the current and voltage of the resistor R, we obtain

$$E = V_D + I_D R$$

In a more meaningful way, this can be rearranged as

$$I_D = \frac{E}{R} - \frac{V_D}{R} \tag{2.2}$$

It is clear that I_D is a function of V_D and can be plotted as a straight line in a figure.

Next, the state of the diode should be determined. In Fig. 2.12 (a), the applied DC supply E is to establish a current in the direction indicated by the arrow in a clockwise direction, which is the same as the arrow in the diode symbol, revealing that the diode is in the "on" state. In other words, the applied voltage results in a forward-bias condition. So, the characteristics, indicated in Fig. 2.12 (b), can be used to solve the network.

Furthermore, the characteristics in Fig. 2.12 (b) also show the relationship between I_D and V_D, which has the same meaning as that of Eq. (2.2). This similarity makes it possible to plot Eq. (2.2) and Fig. 2.12 (b) in one single figure. Then, the intersection of the two curves will determine the solution to the network.

As Eq. (2.2) is a linear relationship, two points can be used to draw the straight line. For simplicity, we set V_D zero to obtain a point on coordinate, that is,

$$I_D|_{V_D=0} = \frac{E}{R}$$

Similarly, we set I_D zero to obtain a point on abscissa, that is,

$$V_D|_{I_D=0} = E$$

So, the two points, $(0, E/R)$ and $(E, 0)$ can be used to plot the straight line, as illustrated in Fig. 2.13. This line is also called a load line.

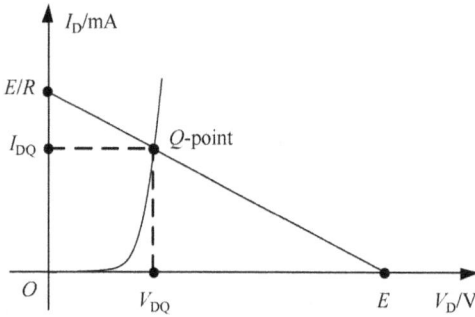

Fig. 2.13: Intersection (Q-point) obtained by the two curves.

Changing the value of R (the load of the circuit) leads to a change in the slope of the load line and a different point of intersection between the device characteristics and the load line. The intersection is the solution to the network, usually called the quiescent point, or simply "Q-point" to show its "motionless, stationary" qualities as only determined by the DC network.

Now the operation point for this circuit is ready. Then, by simply plotting a line down to the abscissa, the diode voltage V_{DQ} can be determined, whereas a horizontal line from intersection to the vertical axis will find the value of I_{DQ}. The current I_{DQ} is actually the current through the resistor R in this series configuration of Fig. 2.12 (a).

Finally, the voltage across resistor R can be obtained by

$$V_R = E - V_{DQ}$$

Solution 2 (mathematical method)
Theoretically, the solution of the network can be obtained by solving the equations

$$\begin{cases} I_D = I_S(e^{V_D/nV_T} - 1) \\ I_D = \frac{E}{R} - \frac{V_D}{R} \end{cases}$$

Note that the first equation contains the exponential function. Some nonlinear method of solving equations would be required. This increases the mathematical complexity greatly and is beyond the coverage of this book.

Now it is obvious that the graphical method introduce above provides a "pictorial" way with a minimum of effort to find the values of V_{DQ} and I_{DQ} for the network.

Solution 3 (equivalent method)
The graphical method has been used to avoid lengthy derivation in solving nonlinear equations. However, sometimes accurate diode characteristics are not available, or the realistic characteristics are slightly different from the theoretical ones. So, the graphical method can be modified by introducing an approximate piecewise-linear

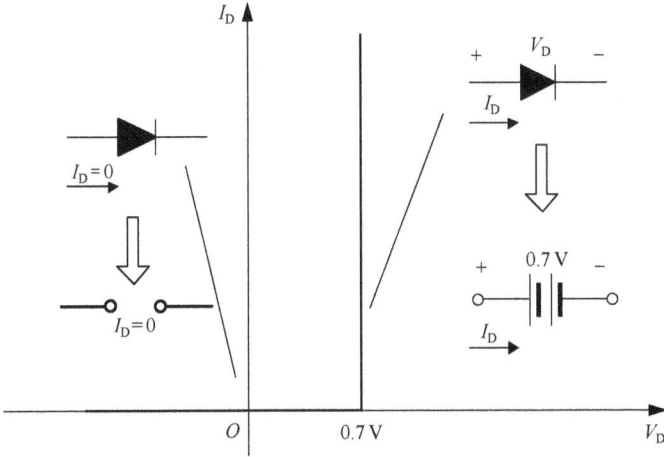

Fig. 2.14: Piecewise-linear equivalent model of a diode.

equivalent model to replace the full characteristics to obtain the parameters of the network. Although the results found by the piecewise-linear equivalent model may not be as accurate as those from the graphical method, they can be regarded as being within the tolerable range of accuracy. More importantly, the process of finding the solution can be much simpler by the piecewise-linear equivalent model.

It is assumed that the forward resistance of the diode is usually so small compared to other series elements of the network that it can be ignored. So, the forward-biased condition can be replaced by a vertical line and the equivalent circuit can be a constant voltage of 0.7 V. (This value is for a silicon diode. For a GaAs diode, this voltage is 1.2 V and for a Ge diode it is 0.3 V. For simplicity, throughout the textbook, 0.7 V will be used.) Also, for the reversed-biased condition, the current I_D is zero and the equivalent circuit can be an open circuit, as shown in Fig. 2.14.

Going back to the network in Fig. 2.12, the state of the diode must be determined first. Which state is the diode in, the forward-biased condition (ON state) or the reversed-biased condition (OFF state)? In general, a diode is in the ON state if the current generated by the applied sources is such that its direction matches that of the arrow in the diode symbol. Otherwise, it is in the OFF state.

To judge the direction of the current, remove the diode and place a resistor in the position, as shown in Fig. 2.15 (a). The resulting direction of I_D is the same as that indicated by the diode symbol, so the diode is in the ON state.

Then, substitute an equivalent circuit, a power source of 0.7 V, for the diode, in the network, and the remaining parameters are the following:

$$V_R = E - 0.7\,\text{V}, \quad I_D = I_R = \frac{V_R}{R}$$

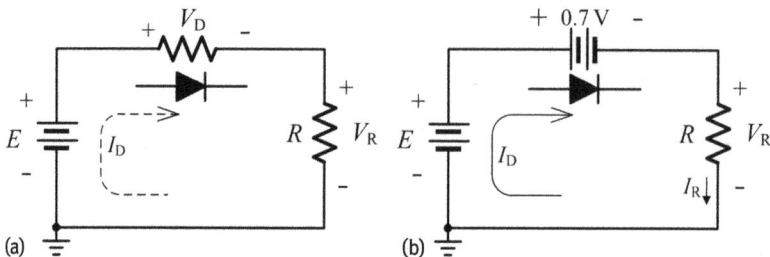

Fig. 2.15: Series diode configuration, (a) Determine the current, (b) Replaced with equivalent circuit.

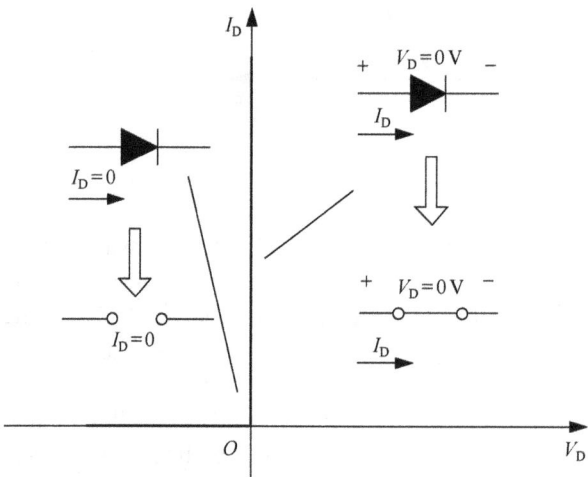

Fig. 2.16: Ideal equivalent model of a diode.

Moreover, in an ideal situation, the piece-wise equivalent circuit can be modified as shown in Fig. 2.16, that is, the voltage drop of the diode in the forward-biased condition has been ignored, and the diode can be replaced with a short circuit.

Logic gates

The diodes can be used to set up a kind of logic circuit, of which the input and output voltages conform to some logic rules. The following analysis will be confined to determining the voltage levels, and thus prove that they follow some logic rules.

First, set up the OR logic of Boolean algebra for the voltages. The 5-V level of voltage corresponds to a "1" for Boolean algebra and the 0-V voltage is assigned a "0". An OR gate is such that the output voltage level will always be a "1" whenever either of the inputs contains "1"; the output is "0" only when both inputs are "0s".

Example 2.2

Analyze the network shown in Fig. 2.17, determine V_o to prove that the network conforms to OR logic, that is, it is a logic OR gate [11].

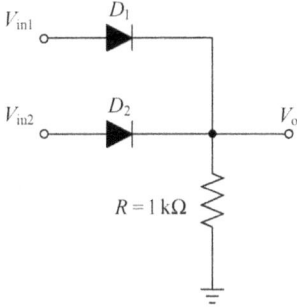

Fig. 2.17: Logic OR gate by diodes.

Solution

Now, the approximate equivalent for a diode is used for the analysis of the OR gate. In general, the state of the diodes can be determined by clarifying the direction and the "pressure" generated by the applied potentials. Through analysis, wrong assumptions will be negated, and correct assumptions will be verified and accepted as truth. There are two inputs, so there is a total of four cases.

Case 1: $V_{in1} = 5$ V, $V_{in2} = 0$ V. In this case, there is only one applied potential, 5 V at input 1, and input 2 is essentially at ground potential. First, assume D_1 is in the OFF state and it is replaced by an open circuit, as shown in Fig. 2.18 (a). Then, any branch in the network except input 1, will be separated from the potential and be at zero volt. So, $V_o = 0$ V.

This means that the negative terminal of D_1 has a lower potential, while the positive terminal has a higher potential. From the diode properties in the previous sections, these states of the two terminals are the same as those of the forward-biased (ON) state. However, D_1 was assumed to be in the OFF state. The contradiction comes

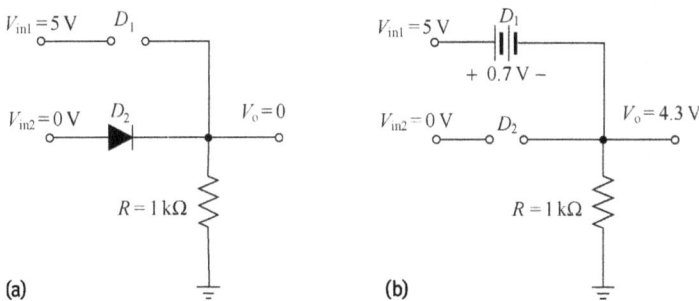

(a)

(b)

Fig. 2.18: Case 1 of Example 2.2, (a) Assumed state of D_1, (b) Correct states of D_1 and D_2.

from the assumption. So, the assumption should be negated, and D_1 is in the ON state. It can be replaced with the 0.7 V potential, as shown in Fig. 2.18 (b).

Then, negative terminals of D_1, D_2 and V_o are all 4.3 V. As the positive terminal of D_2 is at 0 V, the state of D_2 is OFF (it cannot be ON) and can be replaced with an open circuit and redrawn as in Fig. 2.18 (b).

The current of D_1 flows in the direction of the diode symbol (this further proves the assumed ON state of D_1) and is the same as that of the resistor. There is no contradiction and the truth is that D_1 is ON and D_2 is OFF. The output voltage level V_o is 4.3 V and is sufficiently large to be considered logic "1".

Case 2: V_{in1} = 0 V, V_{in2} = 5 V. As the two diodes D_1 and D_2 are in the same status in the network, so Case 2 has the same output as in Case 1.

Case 3: V_{in1} = 5 V, V_{in2} = 5 V. Since the two anodes (positive terminals) of diodes D_1 and D_2, are 5 V, both diodes are in the ON state and the output voltage level V_o is 4.3 V, logic "1", as illustrated in Fig. 2.19.

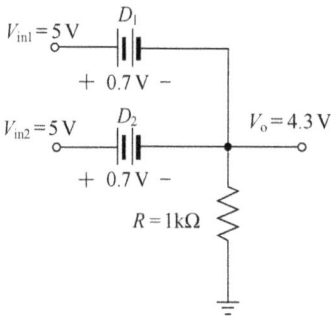

Fig. 2.19: Analysis of Example 2.2 in Case 3.

Case 4: V_{in1} = 0 V, V_{in2} = 0 V. In this case, there is no power fed to the network, so both diodes are in the no-bias condition and the output voltage level V_o is 0 V, logic "0".

The input/output relationship of the four cases are summarized in Tab. 2.1, and it is clear that the output voltage level will always be a "1" whenever either of the inputs contains "1"; the output is "0" only when both inputs are "0", which means that the network conforms to OR logic and the circuit is an OR gate.

Tab. 2.1: Input/output relationship of Example 2.2

V_{in1} ╲ V_{in2}	"1" (5 V)	"0" (0 V)
"1" (5 V)	"1" (4.3 V)	"1" (4.3 V)
"0" (0 V)	"1" (4.3 V)	"0" (0 V)

Example 2.3

Determine the output voltage V_o of the network in Fig. 2.20, proofing that the network conforms to AND logic, that is, it is a logic AND gate [11].

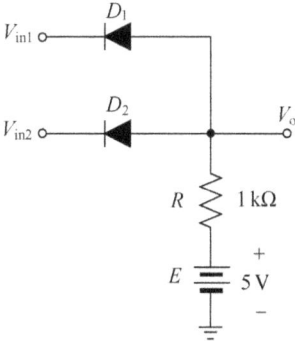

Fig. 2.20: Logic AND gate by diodes.

Solution

Note that in this network an independent source of 5 V appears between the resistor and the ground. As there are two inputs, there is a total of four cases.

Case 1: $V_{in1} = 5$ V, $V_{in2} = 0$ V. With 5 V at the cathode of D_1, the anode cannot be higher than cathode, because all the potentials applied to the network are not higher than 5 V. So, it will be in the no-bias condition or the "OFF" state. If it is in the no-bias condition, then the cathode and anode should have the same voltage, that is, 5 V. So, D_1 can be replaced with a short circuit as shown in Fig. 2.21 (a). Then, the anode of D_2 is at 5 V, while the cathode is at 0 V. So, D_2 is in the "ON" state. Then D_2 can be replaced with a 0.7 V potential, leading to its anode at 0.7 V. This is in conflict with the conclusion from the assumption that the anode of D_2 is at 5 V. Then the assumption that D_1 is in the no-bias condition is wrong. Moreover, only one choice for the state of

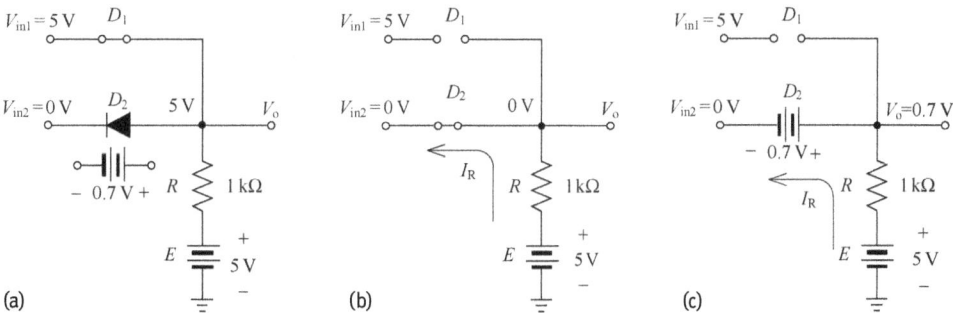

Fig. 2.21: Case 1 of Example 2.3, (a) Assumed state of D_1, (b) Assumed state of D_2, (c) True states of D_1 and D_2.

D_1 remains, that is, it must be in the "OFF" state. Therefore, because the D_2 cathode is grounded, D_2 is assumed to be in the no-bias condition or in the "ON" state.

If D_2 is in the no-bias condition, then the anode is at 0 V, and the resistor will have a current through it, which is the same as the one flowing through D_2, as shown in Fig. 2.21 (b). D_2 has a current, and this is in conflict with its assumed state of no bias.

So, the assumed no-bias condition is wrong, and the true state is "ON", as shown in Fig. 2.21 (c).

So, the output of the network V_o is 0.7 V due to the forward-biased diode D_2. The current I_R will have the direction indicated in Fig. 2.21 (c) and a magnitude equal to

$$I_D = I_R = \frac{E - V_{D2}}{R} = \frac{5\,V - 0.7\,V}{1\,k\Omega} = 4.3\,mA$$

Please note that the output voltage is 0.7 V and is sufficiently small to be considered as the logic "0".

Case 2: $V_{in1} = 0\,V$, $V_{in2} = 5\,V$. As the two diodes, D_1 and D_2, are in the same status in the network, Case 2 has the same output as Case 1.

Case 3: $V_{in1} = 5\,V$, $V_{in2} = 5\,V$. Because all the potentials applied to the network are not higher than 5 V, and both the cathodes of D_1 and D_2 are already at 5 V, the anodes cannot be higher than the cathodes, So, D_1 and D_2 will be in the no-bias condition or in the "OFF" state.

If D_1 and D_2 are in the "OFF" state, as shown in Fig. 2.22 (a), both diodes can be replaced with open circuits and there is no current in the network, leading to 5 V of output V_o.

At the same time, if D_1 and D_2 are in the no-bias condition, then the anodes will be at 5 V, and there is no current through the resistor R, leading to 5 V of output V_o, as shown in Fig. 2.22 (b).

Considering both assumptions, it is clear that both D_1 and D_2 are in the no-bias condition and there is no current flowing in the network. The output V_o is 5 V, logic "1".

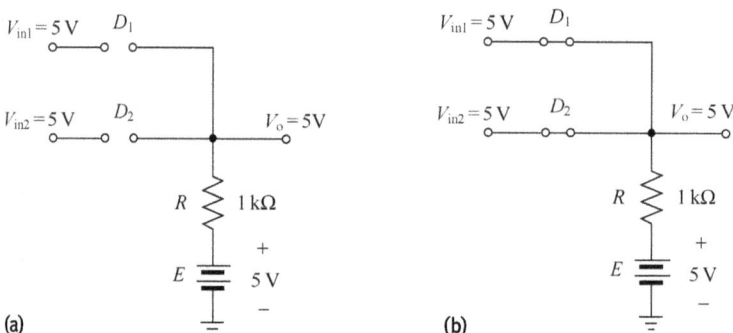

Fig. 2.22: Case 3 of Example 2.3, (a) Assumed states of D_1 and D_2, (b) True states of D_1 and D_2.

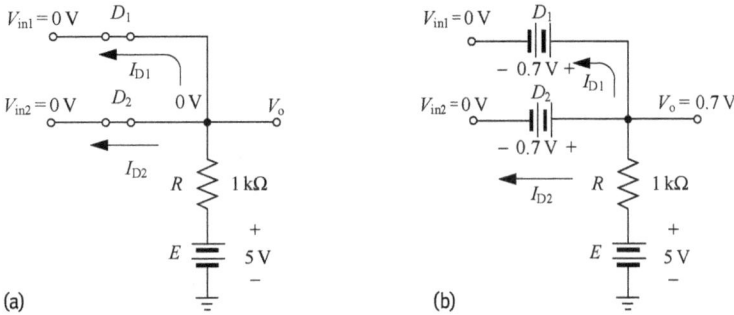

Fig. 2.23: Case 4 of Example 2.3, (a) Assumed states of D_1 and D_2, (b) True states of D_1 and D_2.

Case 4: $V_{in1} = 0\,\text{V}$, $V_{in2} = 0\,\text{V}$. Because both the cathodes are grounded, D_1 and D_2 are assumed to be in the no-bias condition or in the "ON" state.

If D_1 and D_2 are in the no-bias condition, the anodes are at $0\,\text{V}$, and the resistor will have a current, which is shared equally by D_1 and D_2, as shown in Fig. 2.23 (a). This is in conflict with the assumed state of no bias. So, the assumed no-bias condition is wrong, and the true states are "ON", as shown in Fig. 2.23 (b).

So, the output of the network V_o is 0.7 V due to the forward-biased diodes. Also, note that the output voltage is 0.7 V and is sufficiently small to be considered as logic "0".

The input/output relationships of the four cases are summarized in Tab. 2.2, and it is clear that the output voltage level will always be a "0" whenever either of the inputs contains a "0"; the output is "1" only when both inputs are "1", which means that the network conforms to AND logic and the circuit is an AND gate.

Tab. 2.2: Input/output relationship of Example 2.3

V_{in1} \ V_{in2}	"1" (5 V)	"0" (0 V)
"1" (5 V)	"1" (5 V)	"0" (0.7 V)
"0" (0 V)	"0" (0.7 V)	"0" (0.7 V)

Now, the diode application can be expanded to the input of time-varying functions, such as the sinusoidal waveform and the square wave.

Example 2.4

Analyze the network of half-wave rectifier, as shown in Fig. 2.24 (a) [11, 13].

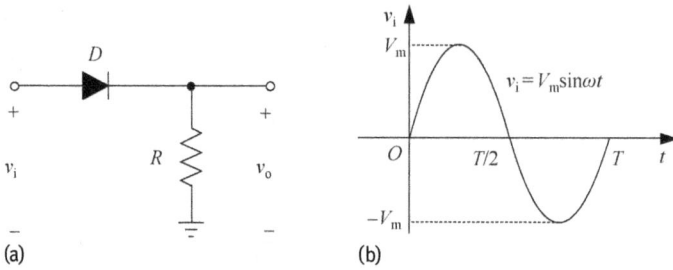

Fig. 2.24: Network and input of Example 2.4, (a) Half-wave rectifier, (b) Sinusoidal waveforms input.

Solution

First, the ideal model of the diode will be assumed to reduce the mathematical complexity. For analysis purposes, one period of the sinusoidal waveform is illustrated in Fig. 2.24 (b).

Also, in one full period, the average value (the algebraic sum of the areas above and below the axis) is zero. The purpose of the rectifier is to generate a waveform V_0 that will have a nonzero average value, which is of particular use in the AC-to-DC conversion process.

During the interval [0, $T/2$], the polarity of the applied voltage v_i is positive, and the current will have the same direction indicated by the arrow of the diode symbol. So, the diode will be in the "ON" state and short-circuit equivalence can be used to replace it, as shown in Fig. 2.25 (a). Then, the terminal of the output voltage is connected directly to the input signal via the short-circuit equivalence of the diode. It is fairly clear that the output signal is an exact copy of the input signal, as shown in Fig. 2.25 (b).

During the interval of [$T/2$, T], the polarity of the applied voltage v_i is negative, and the current may have "pressure" opposite the arrow of the diode symbol. So, the diode will be in the "OFF" state and open-circuit equivalence can be used to replace it, as shown in Fig. 2.26 (a). Then, the terminal of the output voltage is separated from the input signal, and the output voltage is zero, as shown in Fig. 2.26 (b).

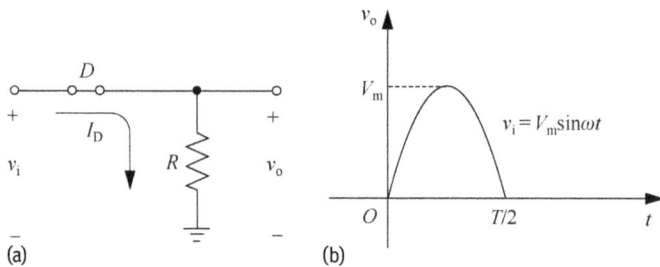

Fig. 2.25: Analysis in the positive half-cycle, (a) Short-circuit equivalence, (b) Output waveform.

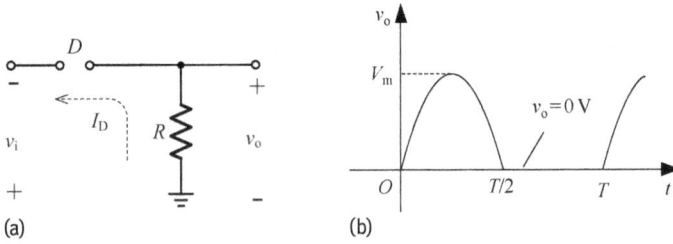

(a)

(b)

Fig. 2.26: Analysis in the negative half-cycle, (a) Open-circuit equivalence, (b) Output waveform.

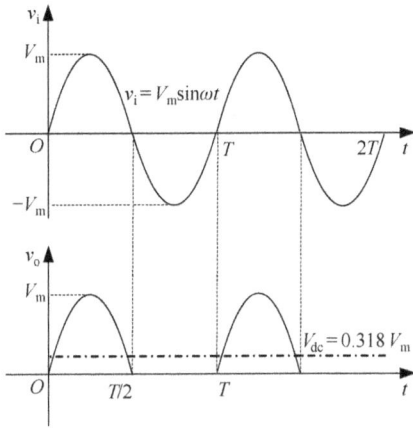

Fig. 2.27: Comparison between input and output waveforms of a half-wave rectifier.

As illustrated in Fig. 2.27, the network removes one half-period of the input signal to establish a nonzero mean value, and this circuit is normally called a half-wave rectifier.

The output signal v_o has a net positive area above the axis, which can be obtained by

$$S_{\text{half}} = \int_0^{T/2} \sin \omega t\, dt$$

$$= \frac{1}{\omega} \cos \omega t \Big|_{T/2}^{0}$$

$$= \frac{1}{\omega} \left(\cos 0 - \cos \omega \frac{T}{2} \right)$$

Substituting the relationship $T = \frac{2\pi}{\omega}$ into it, we get

$$S_{\text{half}} = \frac{1}{\omega}(\cos 0 - \cos \pi) = \frac{2}{\omega}$$

and over a full period, an average value is determined by

$$V_{\text{dc}} = \frac{S_{\text{half}}}{\text{Period}} = \frac{2/\omega}{T} = \frac{2/\omega}{2\pi/\omega} = \frac{1}{\pi} = 0.318$$

A half-wave rectifier can only use 50% of the input power. To improve the efficiency to 100%, bridge configuration with four diodes can be used.

Example 2.5
Analyze the network of a full-wave rectifier, as shown in Fig. 2.28 (a) [11, 13].

Solution
For analysis purposes, one period of the sinusoidal waveform is illustrated in Fig. 2.28 (b). Also, the ideal model of the diode will be used to reduce the mathematical complexity.

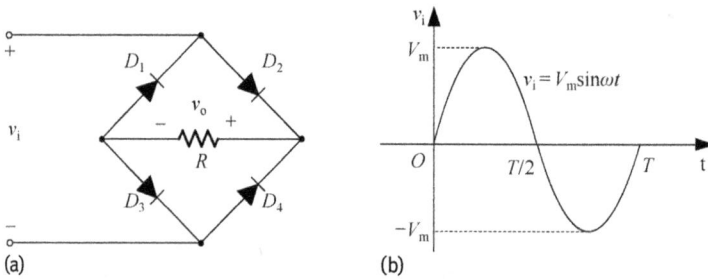

(a) (b)

Fig. 2.28: Full-wave bridge rectifier, (a) Half-wave rectifier, (b) Sinosidal waveforms input.

During the positive half-cycle, the interval of $[0, T/2]$, the polarity of the applied voltage v_i is positive, and the current will have the same direction as indicated by the arrow of the diode symbols of D_2 and D_3. So, the diodes D_2 and D_3 will be in the "ON" state. Similarly, the current will have the opposite direction indicated by the arrow of the diode symbols of D_1 and D_4. So, the diodes of D_1 and D_4 are in the "OFF" state, as illustrated in Fig. 2.29 (a).

Then, D_2, D_3 are replaced with short circuits and D_1, D_4 open circuits. The output terminals are connected directly to those of the input signal via the short-circuit equivalence of the diodes. It is obvious that the output voltage is an exact copy of the input signal, as indicated in Fig. 2.29 (b).

During the negative half-cycle, the interval of $[T/2, T]$, the polarity of the applied voltage v_i is negative, and the current will have the same direction as indicated by the arrow of the diode symbols of D_1 and D_4. So, the diodes D_1 and D_4 will be in the "ON" state. Similarly, the current will have the opposite direction indicated by the arrow of the diode symbols of D_2 and D_3. So, the diodes of D_2 and D_3 are in the "OFF" state, as illustrated in Fig. 2.30 (a).

Then, D_1, D_4 are replaced with short circuits and D_2, D_3 open circuits, as indicated in Fig. 2.30 (b). The satisfactory result is that the polarity across the load resis-

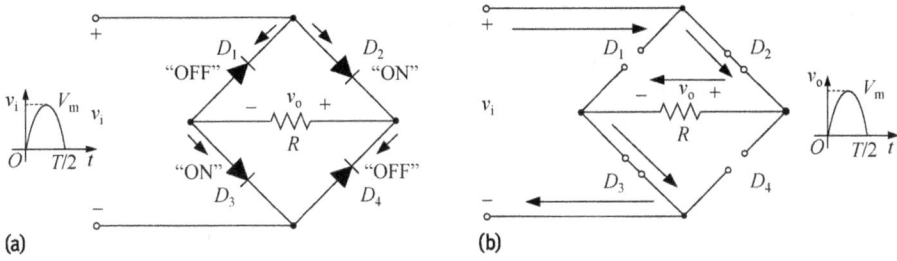

Fig. 2.29: Analysis in the positive half-cycle, (a) Diodes states, (b) Equivalence model.

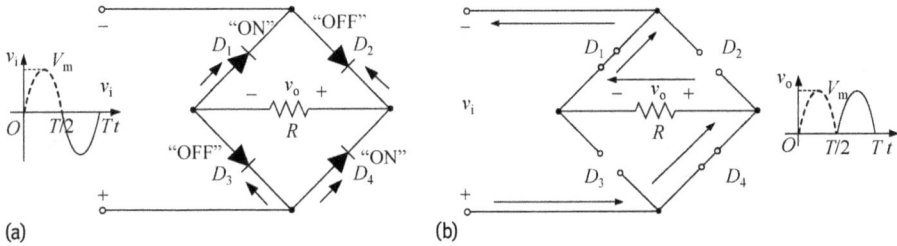

Fig. 2.30: Analysis in the negative half-cycle, (a) Diodes states, (b) Equivalence model.

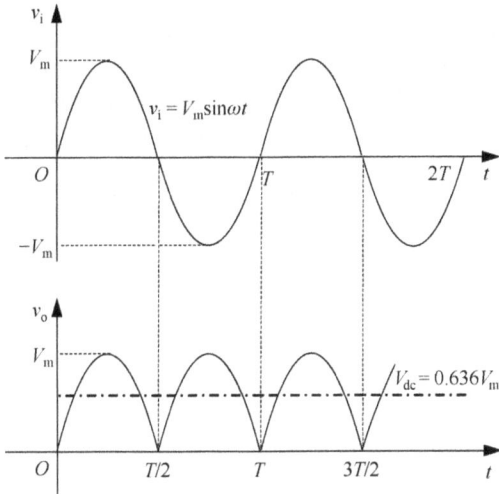

Fig. 2.31: Comparison between input and output waveforms of full-wave rectifier.

tor R is the same as that in the positive half-cycle, leading to a second positive pulse, as shown in Fig. 2.30 (b).

In Fig. 2.31, the input and output waveforms are shown to give a clear comparison. Since during one full cycle, the positive area of output voltage is twice as much as that from the half-wave rectifier, the DC level has been doubled

$$V_{dc} = 0.636\, V_m$$

Please note that in the analysis of both the half- and the full-wave rectifiers, the ideal equivalence models are involved, and the output waveform peaks are as high as the theoretical values. If diode conduction voltage drop is considered, the realistic output peaks will be lower than the theoretical values.

Example 2.6

For the Zener diode network in Fig. 2.32, explain the role which the Zener diode plays while the load resistor R_L is changed to different values [11].

Solution

From Fig. 2.10, the states of the Zener diode are determined by the voltage externally applied to it. Also, from Fig. 2.32, it is clear that the Zener diode is reversed biased in the network and V_Z is an important parameter used to judge the states of the Zener diode. So, first determine the voltage V_L across the Zener diode, as shown in Fig. 2.33.

From the basic knowledge of circuits, it is clear that

$$V_L = \frac{R_L}{R_L + R} E$$

Now, the plot of the relationship between V_L and R_L without consideration of the Zener diode is shown in Fig. 2.34 (a). V_L will increase from zero and approach the voltage of the power supply with increasing R_L values. However, with consideration of the Zener diode, which has a Zener potential V_Z of 10 V, the curve will change. First, while V_L is lower than 10 V, the diode is in the "OFF" state and the curve of $V_L \sim R_L$ is same as that in Fig. 2.34 (a). However, the increase of V_L will be interrupted when

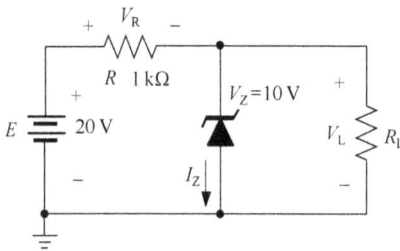

Fig. 2.32: Network with the Zener diode for Example 2.6.

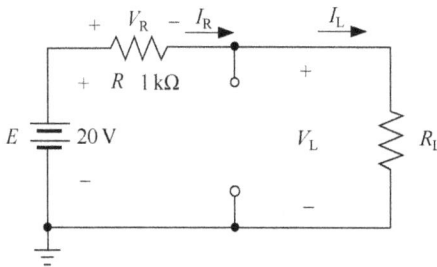

Fig. 2.33: Determining V_L to judge the state of the Zener diode.

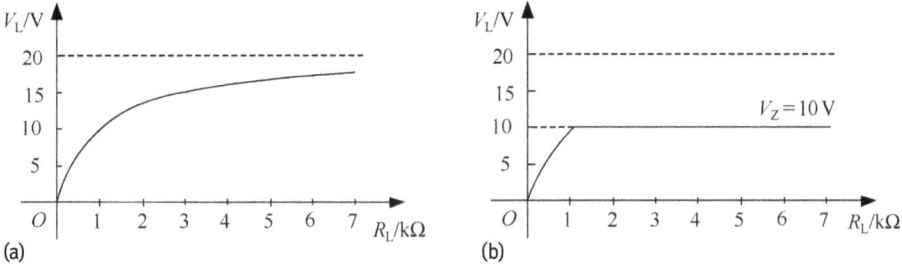

(a)   (b)

Fig. 2.34: Relationship between V_L and R_L, (a) Without consideration of V_Z, (b) With consideration of V_Z.

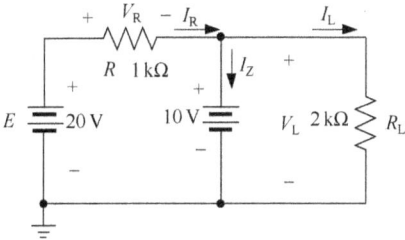

Fig. 2.35: Equivalent circuit when the Zener diode is in the Zener region.

V_L reaches or tries to exceed V_Z. The state of the Zener diode is in the Zener region, as shown in Fig. 2.10. So, the voltage will be limited to V_Z, as illustrated in Fig. 2.34 (b), and the current is determined by other part of the network.

Now, the currents in the network can be obtained from Fig. 2.35, when R_L is 2 kΩ, for example.

$$V_R = E - V_Z = 20\,\text{V} - 10\,\text{V} = 10\,\text{V}$$

$$I_R = \frac{V_R}{R} = \frac{10\,\text{V}}{1\,\text{k}\Omega} = 0.01\,\text{A}$$

$$I_L = \frac{V_L}{R_L} = \frac{10\,\text{V}}{2\,\text{k}\Omega} = 0.005\,\text{A}$$

So,

$$I_Z = I_R - I_L = 0.01\,\text{A} - 0.005\,\text{A} = 0.005\,\text{A}$$

The power dissipated by the Zener diode is

$$P_Z = V_Z I Z = 10\,\text{V} \cdot 5\,\text{mA} = 50\,\text{mW}$$

which should be within the maximum value of the device.

The role played by the Zener diode in this example is that it can limit the voltage V_L when the load resistor R_L is too large, for some kind of protection purpose.

2.7 Chapter summary

Important concepts and conclusions:
(1) A *p-n* junction is a piece of semiconductor material in which part of the material is *p*-type and part is *n*-type. See Section 2.1.
(2) A semiconductor diode is a device containing a *p-n* junction mounted in a package with two connecting leads. See Section 2.2.
(3) There are three biasing conditions for diodes: the no-bias condition, the reverse-bias condition and the forward-bias condition. See Section 2.2.
(4) The equivalent model for a reverse-biased diode is an open circuit. The equivalent model for a forward-biased diode is a 0.7 V voltage drop or a short circuit. See Section 2.2.
(5) The reverse-bias voltage that leads to a dramatic increase in characteristics for a diode is called the Zener potential and is designated as V_Z. Diodes employing this unique portion of the characteristic of a *p-n* junction are called Zener diodes. See Section 2.5.
(6) The equivalent model of Zener diodes is just a DC battery of V_Z V. See Section 2.5.
(7) A piecewise-linear equivalent model of a diode can be used to analyze circuits. See Section 2.6.
(8) In order to determine the state of a Zener diode in a circuit, first remove it from the position and evaluate the voltage between the two points originally connected to the Zener diode pins. If the voltage is more than the Zener potential and has the correct polarity, the Zener diode is in the Zener region. Otherwise, it is only in reverse-biased condition. See Example 2.6 in Section 2.6.

2.8 Questions

Q2.1: Give your explanation of what happens when *n*- and *p*-type materials are joined together.

Q2.2: What has been depleted in the depletion region?

Q2.3: When *n*- and *p*-type materials are joined together, an internal electrical field is set up near the border. Can this field generate a current if the two outer sides of the materials are connected by a wire?

Q2.4: Before the semiconductor diode was invented, engineers were already using diodes. What were those types of diodes? Find the answer on the web and see how those devices were developed based on different principles.

Q2.5: Summarize the properties of diodes in no bias, forward bias and reversed bias conditions.

Q2.6: The diode characteristics are shown in Fig. 2.5. Please calculate the resistance of the diode from the figure. What conclusion can be reached?

Q2.7: Eq. (2.1) shows the relationship between the voltage and the current of the diode. Can the relationship be used to fulfill an exponential or logarithmic function? Read Section 5.6 for help.

Q2.8: What is the main difference between the Schottky diode and general diodes? Explain why the Schottky diode is better than general diodes in rectifiers.

Q2.9: List the advantages of LEDs when used as lighting devices. Also list several LED applications that have appeared in recent years in daily life or industry.

Q2.10: What is the Zener region? What is the difference in the ways that Zener diodes and general diodes are used? What is the main purpose for using the Zener diode?

Q2.11: How are the characteristics of the diode in Fig. 2.5 approximated as in Fig. 2.14? What is the condition for the simplification?

Q2.12: A diode circuit is shown in Fig. 2.36, $v_i = 6 \sin 2\pi ft$ V. Plot the output v_o in conditions when diode $V_T = 0$ or $V_T = 0.7$ V with $E = 1.5$ V or $E = -1.5$ V.

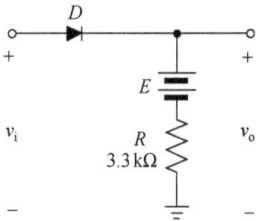

Fig. 2.36: Network of Q2.12.

3 Bipolar junction transistors

In electronics, a vacuum tube, or an electron tube, also called a tube or a valve by people from different regions, is a device that controls the electric current between electrodes (anode and cathode) in an evacuated glass envelope. Vacuum tubes mostly work on the principles of thermionic emission of electrons from a cathode heated by a filament [20]. As an example, a vacuum tube is shown in Fig. 3.1. Its glass envelope contains low-pressure gas with electrodes inside it, and the upper part of it has a silver coating on the internal surface.

Fig. 3.1: A vacuum tube.

In 1904 vacuum tubes were invented and were fundamental components for electronics throughout the first half of the twentieth century, which saw the widespread of radio and television broadcasting, radar systems, sound amplification, recording and duplication, telecommunication networks, analog and digital computers and industrial process control. It was the invention of the vacuum tube that made these technologies practical, accelerated industrial development and improved humans' daily lives.

However, in the 1940s, the invention of semiconductor devices made it possible to produce solid-state devices, which are smaller, cheaper, more lightweight, more re-

https://doi.org/10.1515/9783110593860-003

liable, more efficient, more durable and more convenient to use than vacuum tubes. Hence, from the mid-1950s solid-state devices, such as transistors, gradually superseded tubes. Only a few high-power applications, such as the magnetron used in microwave ovens and certain high-frequency amplifiers, remain for vacuum tubes. Now, the cathode-ray tube (CRT) for televisions and video monitors remain in some people's childhood memories, for these were replaced by LED or LCD (liquid-crystal display) display with the coming of the twenty-first century.

Starting from the semiconductor diode, which contains one p-n junction, now transistors with two p-n junctions possess more valuable properties and characteristics.

3.1 Fundamentals of transistors

3.1.1 Structure of transistors

The transistor is a three-layered semiconductor device consisting of either two p- and one n-type layers of material, which is called a pnp transistor, or two n- and one p-type layers of material, which is called an npn transistor. Both are shown in Fig. 3.2 with the DC biasing necessary to establish the proper region of operation for AC amplification and will be discussed in detail in later sections.

This three-terminal device is called the bipolar junction transistor, or simply BJT. The term bipolar comes from the fact that holes and electrons are involved in the injection process into the oppositely polarized material. If only one type of carrier is employed (electron or hole), it is referred to as a unipolar device, which will be introduced in subsequent chapter. For the biasing shown in Fig. 3.2, the terminals are indicated by the capital letters E for emitter, C for collector and B for base [1–3].

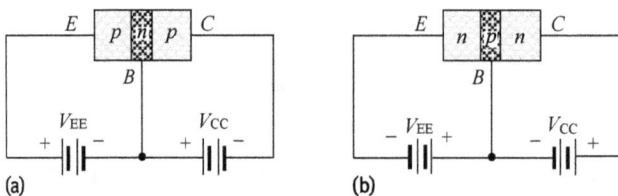

Fig. 3.2: Structure of transistors with proper biasing, (a) *pnp* type, (b) *npn* type.

Normally, the emitter layer is heavily doped, and the base and the collector are lightly doped. Moreover, the doping of the sandwiched layer is less than that of the outer layers. Also, the outer layers have widths that are much greater than the sandwiched layer. Taking the transistor in Fig. 3.2 as an example, the total width of the three layers is 100 times wider than that of the center layer.

3.1.2 Transistor operation

Now, using the *npn* transistor as an example, the basic operation of the transistor will be described. The operation of the *pnp* transistor is exactly the same if the roles played by the electrons and holes are interchanged. In Fig. 3.3 (a), the focus is the *b-e* (base-to-emitter) junction of the *npn* transistor, ignoring the *c-b* (collector-to-base) junction temporarily. It is quite clear that this situation is the same as the forward-biased condition of diode described in Chapter 1. Due to the externally applied biasing voltage V_{EE}, the depletion region has been squeezed in width, resulting in a strong flow of majority carriers (positive particles) from the *p*-type material to the *n*-type material.

On the other hand, consider the *c-b* (collector-to-base) junction as shown in Fig. 3.3 (b), ignoring the *b-e* (base-to-emitter) junction temporarily. This situation is the same as that of the reverse-biased condition of diode described in Chapter 1. Due to the externally applied biasing voltage V_{CC}, the depletion region has been widened, resulting in zero majority-carrier flow and a very small minority-carrier (positive particles) flow. Therefore, the basic biasing for transistor is that one *p-n* junction is forward-biased, whereas the other is reverse biased [9, 10].

Now, consider the current through the three terminals of the transistor. In Fig. 3.4, both biasing voltages, V_{CC} and V_{EE}, are shown, with the resulting majority and minority carrier flows indicated through the layers. The depletion region with the forward-biased condition looks thinner and the one with reverse-biased condition looks wider [11].

Since sandwiched *p*-type material is very thin with low conductivity, a very small number of majority carriers will flow from the base terminal to the emitter terminal. So the magnitude of the base current I_B is typically very small, on the order of microamperes [12].

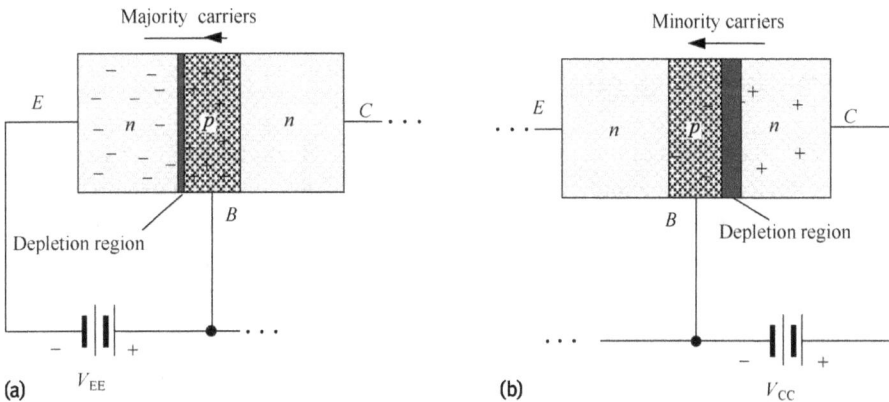

Fig. 3.3: Transistor operation, (a) Forward-biased *b-e* junction, (b) Reverse-biased *c-b* junction.

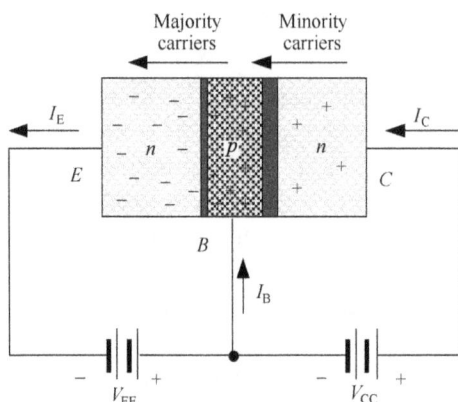

Fig. 3.4: Current flows through the terminals of an *npn* transistor.

In the reverse-biased *c-b* junction, due to the externally applied voltage V_{CC}, minority carriers (positive particles) in the *n*-type layer have been injected into the sandwiched *p*-type layer as majority carriers. These carriers form the collector current, I_C. Combing I_C with majority carriers originally from the sandwiched *p*-type layer, a larger number of majority carriers will diffuse across *b-e* junction, which is forward-biased with thin depletion region, resulting in heavy current flow through the emitter terminal, indicated as I_E. Normally, emitter and collector currents, I_E and I_C, are in the level of milliamperes [13, 16].

Applying Kirchhoff's current law to the *npn* transistor in Fig. 3.4, we obtain

$$I_E = I_C + I_B \tag{3.1}$$

This shows that the emitter current I_E is the sum of the collector and the base currents, I_C and I_B. Although obtained from the *npn* transistor, this relationship is also true for the *pnp* transistor with the directions of the currents reversed.

3.2 Transistor configurations

The electrical symbols of *npn* and *pnp* transistors are illustrated in Fig. 3.5 as three-terminal devices. Throughout this textbook, all current directions will refer to the flow of conventional positive particles (holes) rather than electrons. Moreover, the arrows in all electronics symbols show the directions of currents. This is the same as the arrow in the diode symbol in Chapter 1. For the bipolar junction transistor, the arrow in the graphic symbol defines the direction of the emitter current (conventional flow) through the device. Thus, the directions of the arrows are used to distinguish different types of BJT, as shown in Fig. 3.5 [1].

In the following chapters, these symbols will be used together with other electrical elements to analyze the properties of bipolar junction transistors. As a three-terminal device, each terminal of BJT can be used simultaneously for both input and output

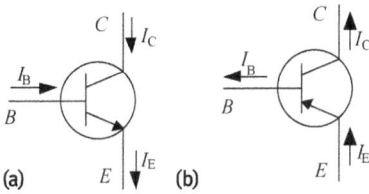

Fig. 3.5: Electrical symbols of transistors, (a) *npn* type, (b) *pnp* type.

loops in the network, that is, "common to both input and output". So, there can be three types of configuration, common-base, common-collector and common-emitter configurations. The following sections are devoted to the analysis of these three configurations. Also, note that all three configurations are only for DC equivalent circuits; realistic circuits will be introduced in later sections.

3.2.1 Common-base configuration

The terminology "common-base", is derived from the fact that the base terminal is common to both the input and output loops of the configuration, as shown in Fig. 3.6 [3, 11].

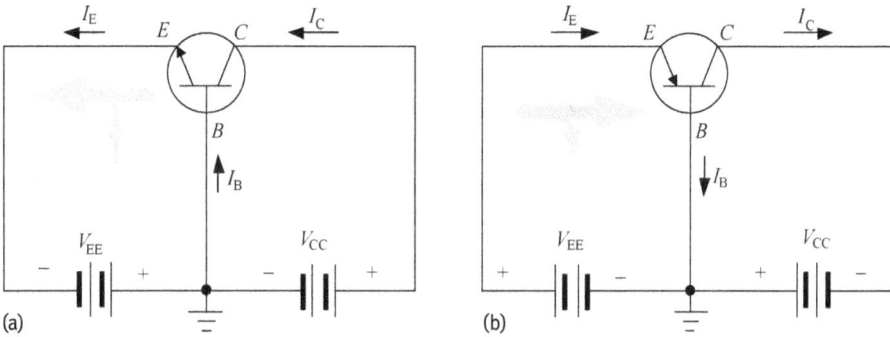

Fig. 3.6: Common-base configurations, (a) *npn* transistor, (b) *pnp* transistor.

All the current directions appearing in Fig. 3.6 are actual directions of conventional flow, and the applied biasing voltages, V_{EE} and V_{CC}, are such as to set up current in the direction indicated for each branch. In other words, the direction of I_E conforms to the polarity of V_{EE} and I_C to V_{CC}. The direction of I_B can be determined from those of I_E and I_C. Note that Eq. (3.1) still holds for both *npn* and *pnp* transistors.

Normally, two sets of characteristics are used to fully describe the behavior of common-base configuration: one for input parameters and the other for the output loop. As shown in Fig. 3.7, the input set for the common-base configuration relates an input current (I_E) to an input voltage (V_{BE}) for different levels of output voltage (V_{CB}).

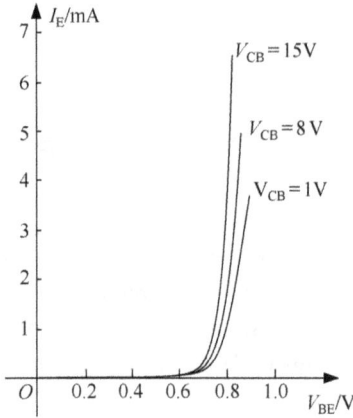

Fig. 3.7: Input characteristics for common-base configuration.

The output set shows the relationship between an output current (I_C) and an output voltage (V_{CB}) for different levels of input current (I_E), as shown in Fig. 3.8. Moreover, the output characteristics have three basic regions, which are indicated by different shading patterns: the active, cutoff and saturation regions. The active region is the portion within which the relationship between I_C and I_E is nearly linear. This part is normally employed for linear amplifiers, which will be discussed extensively later. More precisely speaking, the active region corresponds to the condition that the b-e junction is forward biased, whereas the c-b junction is reverse biased. This part will be used to set up the DC operation point for the amplifier to receive input signals, leading to a fluctuating (active) AC operation point [3, 11].

In the active region, V_{CB} shows little effect on the collector current I_C. As I_E increases above zero, I_C increases to a magnitude almost equal to that of I_E. The curves clearly illustrate an approximation relationship between I_E and I_C in the active region:

$$I_E \approx I_C \tag{3.2}$$

The cutoff region is defined as the portion within which the I_C is below zero, as shown in Fig. 3.8. In addition, the cutoff region corresponds to the condition that the b-e and c-b junctions are both reverse biased, that is, the transistor has been totally cut off. This is the reason that this part got its name.

The saturation region is the part of the characteristics to the left of $V_{CB} = 0$ V. To more clearly show the dramatic change in characteristics in this region, the scale of abscissa in this region is expanded. It can be seen that the exponentially increased I_C appears as the voltage V_{CB} increases toward zero. The saturation region corresponds to the condition that both the b-e and c-b junctions are forward biased.

More importantly, in the input characteristics of Fig. 3.7, it can be seen that for constant values of V_{CB}, as V_{BE} increases, I_E increases in a manner that looks like the diode characteristics of Fig. 2.5. In fact, increasing levels of V_{CB} have little effect on the characteristics, which is the property indicated in Fig. 3.8. So, by ignoring the

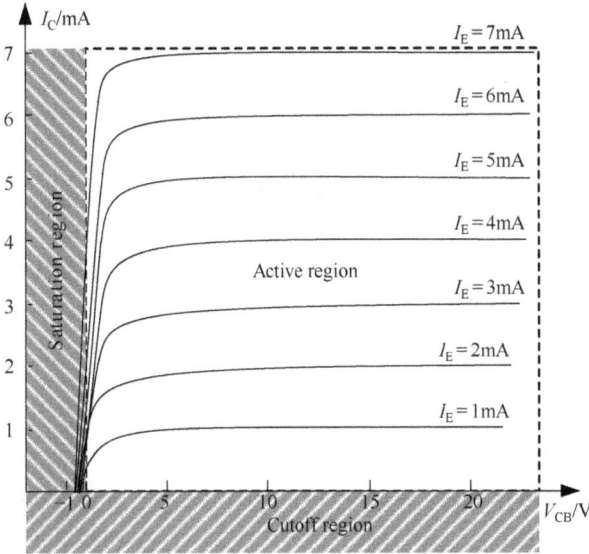

Fig. 3.8: Output characteristics for common-base configuration.

changes in V_{CB}, an approximation of the input characteristics can be drawn as shown in Fig. 3.9 (a) [3, 11].

If line segments are used to replace the curve, the piecewise-linear approximation results as shown in Fig. 3.9 (b). This means that the *b-e* junction has a constant resistance (constant slope of the line segment) when V_{BE} is above 0.7 V.

One step further, by ignoring the slope of the curve, i.e., assuming that the resistance of the *b-e* junction is zero, Fig. 3.9 (c) results. For the analysis of transistor

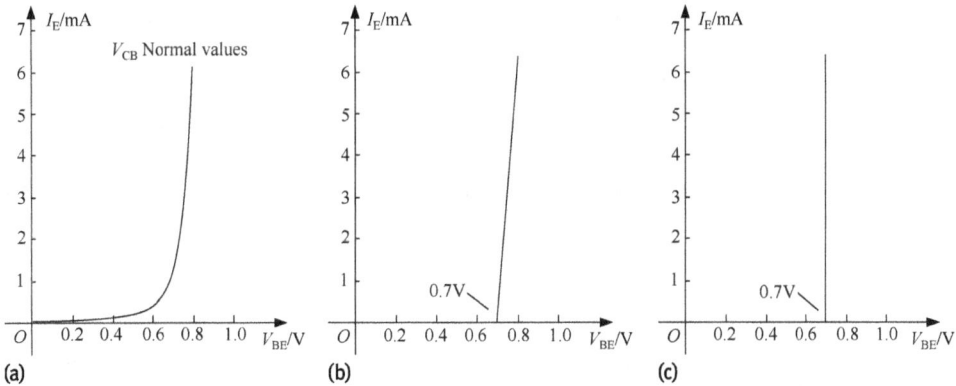

Fig. 3.9: Approximation of the input characteristics, (a) Ignoring the changes in V_{CB}, (b) Piecewise-linear curve, (c) Ignoring resistance of *b-e* junction.

networks throughout this textbook, this will be employed extensively. In other words, once a transistor is set in the active region, the b-e junction is in the "on" state, and the b-e voltage will be regarded as

$$V_{BE} = 0.7 \text{ V} \tag{3.3}$$

With Eq. (3.3), ignoring the influence of V_{CB} and resistance of the b-e junction, a convenient approximation model is obtained to analyze BJT networks with less involvement of trivial parameters and with a sufficiently good level of accuracy. Note that once the BJT is in the "on" or active state, V_{BE} will be 0.7 V regardless of the emitter current I_E, which is controlled by the external network.

Alpha (α)

Because both I_C and I_E are formed by the majority carriers shown in Fig. 3.4, their magnitudes are related by a parameter, *alpha* (α), and defined by the following equation [3, 11]:

$$\alpha \triangleq \frac{I_C}{I_E} \tag{3.4}$$

where I_C and I_E are the levels of current at the instant point of operation. This parameter is for DC mode. From Fig. 3.8, it can be seen that $\alpha \approx 1$. For practical devices, its value typically ranges from 0.90 to 0.998, with most of them almost reaching 1. Normally, Eq. (3.4) is used in the following form,

$$I_C = \alpha I_E \tag{3.5}$$

3.2.2 Common-emitter configuration

By setting the emitter for both the input and output terminals, i.e., common to both the base and collector terminals, the common-emitter configuration is obtained, as illustrated in Fig. 3.10 with *pnp* and *npn* transistors. It is the most widely used transistor configuration in practical applications.

In Fig. 3.10, I_E, I_C and I_B are shown in their actual conventional current directions. The current relations obtained earlier for the common-base configuration are still applicable, although the transistor configuration has changed, that is, $I_E = I_C + I_B$ and $I_C = \alpha I_E$.

Again, input and output characteristics are necessary to explain the behavior of the common-emitter configuration. The input characteristics are a plot of the input current (I_B) versus the input voltage (V_{BE}) for a series of values of output voltage (V_{CE}), as shown in Fig. 3.11 (a). Note that the magnitude of I_B is in microamperes, compared to the milliamperes of I_C [3, 11].

For the common-emitter configuration, the input set of characteristics can also be approximated by a straight-line equivalent that resulted in $V_{BE} = 0.7$ V for any level

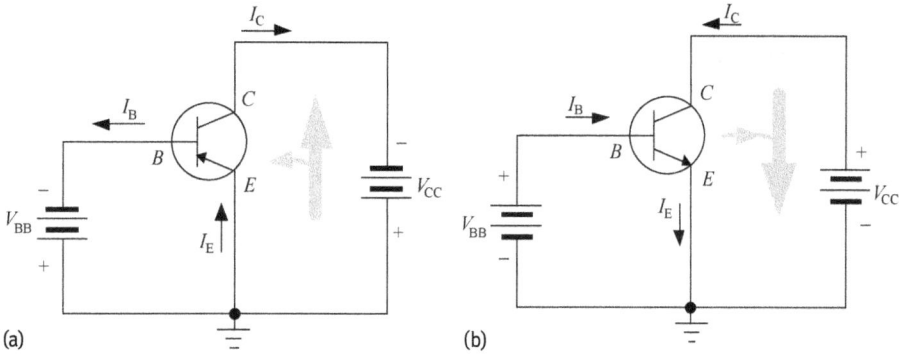

Fig. 3.10: Common-emitter configuration, (a) *pnp* transistor, (b) *npn* transistor.

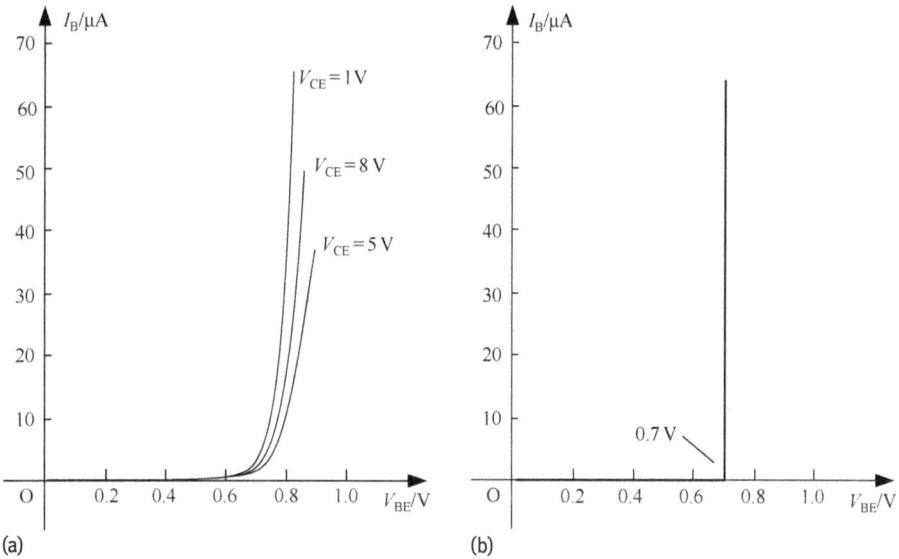

Fig. 3.11: Common-emitter configuration, (a) Input characteristics, (b) Straight-line equivalent.

of I_B greater than 0 mA. In this case, the transistor is in the "ON" state or the active region defined by the output characteristics.

The output characteristics for the common-emitter configuration are a plot of the output current (I_C) versus the output voltage (V_{CE}) for a series of values of input current (I_B), as shown in Fig. 3.12. Consider also that the curves of I_B are not as horizontal as those obtained for I_E in the common-base configuration (illustrated in Fig. 3.8), meaning that the collector-to-emitter voltage V_{CE} will have a greater effect on the magnitude of the collector current I_C.

For the common-emitter configuration, the active region is that part of the upper-right quadrant that possesses the greatest linearity, i.e., the portion within which the

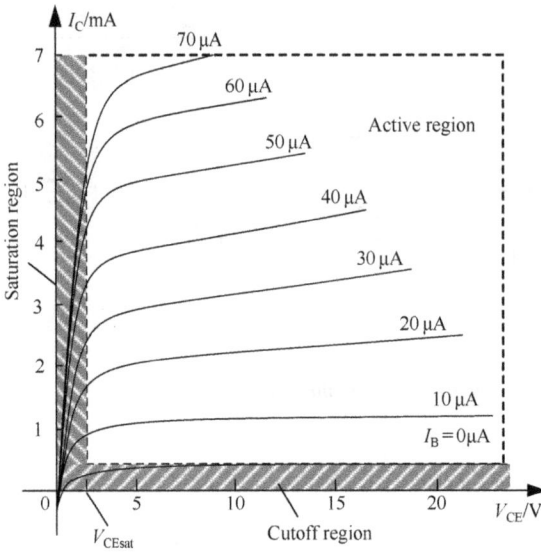

Fig. 3.12: Output characteristics for common-emitter configuration.

curves for I_B are nearly straight and uniformly spaced. In Fig. 3.12, this region has greater V_{CE} than V_{CEsat} and larger I_B than zero. The active region of the common-emitter configuration can be used for current, voltage and power amplification. In the active region of a common-emitter amplifier, the *b-e* junction is forward biased, whereas the *c-b* junction is reverse biased.

For the common-emitter configuration, the region to the left of V_{CEsat} is called the saturation region. The region below $I_B = 0$ is simply defined as the cutoff region. Both the saturation and the cutoff region should be avoided if a linear output signal is required. However, they can be employed as switches in logic gates for digital circuitry.

Beta (β)

In the DC mode a quantity called beta (β) relates the levels of I_C and I_B. It is defined as

$$\beta \triangleq \frac{I_C}{I_B} \tag{3.6}$$

where I_C and I_B are determined at a particular operating point on the characteristics, and this parameter is for the DC mode. For practical transistors, the level of β covers roughly from 50 to more than 400, with most in the midrange. For a transistor with a β of 200, the collector current I_C is 200 times the level of the base current I_B. Normally, Eq. (3.6) is used in the following form, that is,

$$I_C = \beta I_B \tag{3.7}$$

3.2.3 Common-collector configuration

The last transistor configuration is the common-collector configuration, shown in Fig. 3.13 with the proper DC biasing and current indications.

Rather than for amplification purposes, the common-collector configuration is employed mainly for impedance-matching purposes due to the fact that it has a high input impedance (which will be a weak load to previous stage if in a multi-stage system) and low output impedance (strong driving capability), which is opposite to that of the common-base and common-emitter configurations. Moreover, the common-collector configuration can be redrawn in the form shown in Fig. 3.14 with the load resistor R_L connected between emitter and ground. Even though the transistor is connected in a manner similar to the common-emitter configuration, the collector is wired to AC ground, and an input signal is fed from the base terminal. This configuration will be examined in detail in later sections when we discuss DC and AC analysis.

For practical design purposes, the parameters for the common-collector can be obtained from the common-emitter characteristics. The input current I_B is the same for both common-emitter and common-collector characteristics. The output characteristics of the common-collector configuration are a plot of I_E versus V_{EC} for a range of values of I_B. The abscissa for the common-collector configuration is obtained by

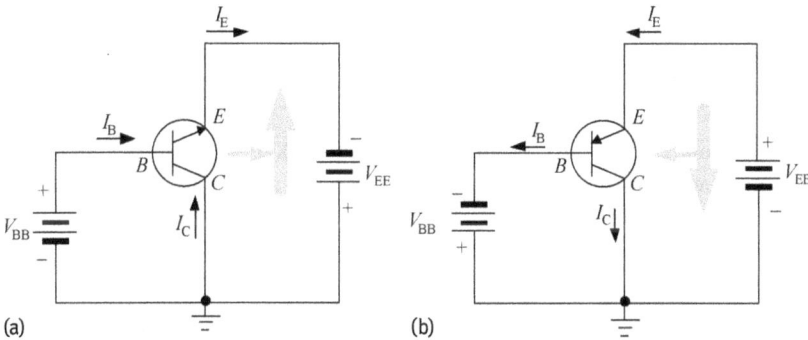

Fig. 3.13: Common-collector configuration, (a) *npn* transistor, (b) *pnp* transistor.

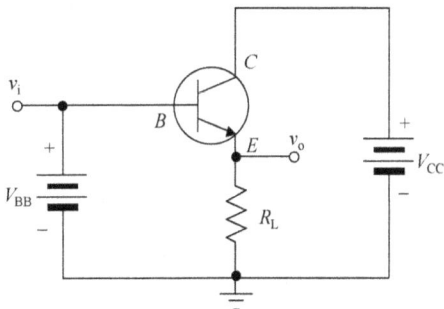

Fig. 3.14: Common-collector configuration used for impedance-matching purposes.

simply changing the sign of V_{CE} of the common-emitter characteristics. Further, it is hardly noticeable to change the vertical coordinate of I_C of the common-emitter characteristics to I_E because of the fact that $I_E \approx I_C$. For the input characteristics of the common-collector configuration, the common-emitter input characteristics are sufficient to provide necessary information [3, 11].

3.3 BJT DC biasing circuits

Actually, in the transistor network, there exist two components: the DC portion and the AC portion. The analysis and synthesis of transistor amplification circuits needs both the DC and the AC response of the system. The amplification of input signals is actually the result of transferring the energy from the biasing DC supplies to the AC output signal by transistors. Fortunately, the superposition theorem is applicable, and the examination of the DC conditions can be totally separated from the AC response. However, it should be borne in mind that the DC levels will also affect the AC response, and vice versa.

Before the introduction of biasing circuits, it is more important to explain the purpose of biasing circuits, that is, to set up a proper DC operation point in a suitable position on the transistor characteristics.

3.3.1 Operating point

The term biasing appearing in the context of this textbook means the process of the application of DC voltages and currents with fixed values in transistor amplification circuits, which results in a point on the characteristics called the operation point or the quiescent point (shortened as Q-point). It is this point defining the region that will be employed for amplification of the applied AC signal, which will make the operation point fluctuate around the original position. Therefore, the Q-point is the basis for the amplification process and is still, motionless, quiet and inactive before the AC signal is applied.

In a transistor amplifier, besides the currents and voltages, the DC level of operation is affected by a number of factors, especially the position of possible operating points on the transistor characteristics. Figure 3.15 shows general transistor output characteristics with four possible operating points indicated, each of which has its coordinates (V_{CE}, I_C) [1, 11].

In Fig. 3.15, the portion at the lower end of the scale is the cutoff region, defined by $I_B \leq 0\,\mu A$ and the saturation region defined by $V_{CE} \leq V_{CEsat}$. Both regions should be avoided when setting up an operation point for amplification purposes.

The maximum ratings are also illustrated on the characteristics by a horizontal dashed line for the maximum collector current I_{Cmax} and a vertical dashed line at the

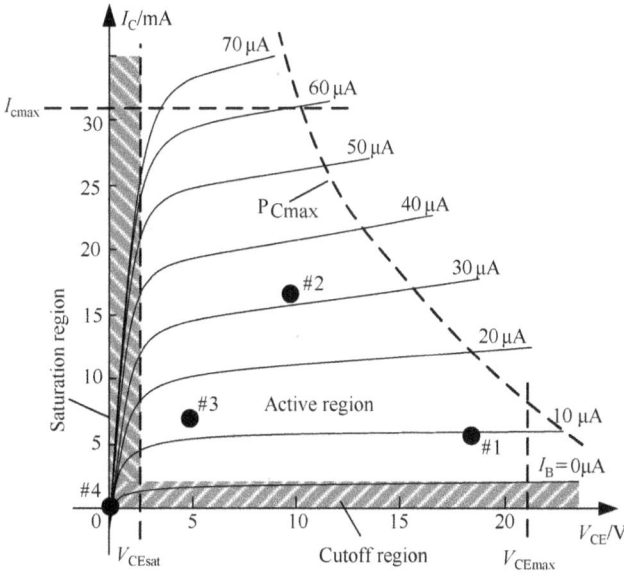

Fig. 3.15: Optional operating points on the transistor characteristics.

maximum collector-to-emitter voltage V_{CEmax}. Also in Fig. 3.15, the maximum power constraint is defined by the curve P_{Cmax}. These dashed lines reduce the area of the active region with which the operation point should be set. If a transistor is biased to operate outside these maximum limits, the result would be either a considerable shortening of the device's lifetime or a direct destruction of the device.

Therefore, an operation point at any position within the active region can be set up, such as the points #1 to #3 and #4 (not in the active region), shown in Fig. 3.15.

Like point #4 in Fig. 3.15, it is actually in no-bias condition and the transistor would be totally turned off. Obviously, there is zero current through the transistor and also no voltage across it. So, when the input AC signal is applied, which contains positive and negative sway, the negative portion of the input would be cut off. For small signal amplification purposes, point #4 is not suitable.

For point #3, it has current and voltage but not far enough from origin. If a signal with large swing is applied, point #3 will fluctuate, leading to negative excursions of the input signal falling into cutoff or saturation region. Also there exists some concern about the nonlinearities introduced by the nonuniform space between I_B curves in this region. So point #3 is still not suitable.

For point #1, it is set near the maximum voltage and power level. The output voltage swing in the positive direction is probably limited, leading to wave shape distortion or device damage. So, point #1 is not suitable either.

Point #2 would be better. It possesses sufficient room for I_C, V_{CE} and power limitation for some positive and negative swing of the output signal. Also, this part of active region will give better linearity to the input signal due to the more equally spaced I_B curves. So, point #2 is more suitable for linear amplification.

In summary, for the transistor to be biased in the active region for linear amplification, the following must be satisfied [11]:

(1) The *b-e* junction must be forward biased, with the forward-bias voltage being about 0.7 V.

(2) The *b-c* junction must be reverse biased, with the reverse-bias voltage of any value not being beyond the maximum limitation.

(3) The selection of the *Q*-point is normally near the center of the active region of the characteristics, keeping sufficient room for the swing of the input signal.

3.3.2 Biasing voltages

As a preparatory step, biasing with correct DC voltage polarities will be discussed first.

The proper steps for biasing a common-base configuration are illustrated in Fig. 3.16. First, determine the direction of I_E from the arrow of the *npn* transistor symbol. Then, the polarity of V_{EE} should be consistent with the direction of I_E, as shown in Fig. 3.16 (b). Assume that $I_B \approx 0$ and $I_C \approx I_E$. So, the polarity of V_{CC} should support the direction of I_C, as illustrated in Fig. 3.16 (c). For the *pnp* transistor, the directions of currents and polarities of voltages should be reversed.

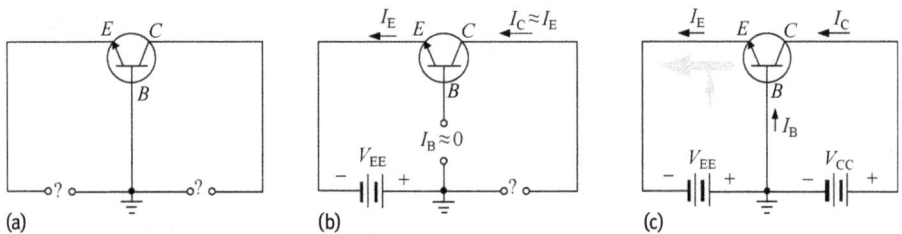

Fig. 3.16: Steps to determine biasing for common-base configuration, (a) Step 1, (b) Step 2, (c) Step 3.

For the common-emitter configuration, the proper steps of biasing are similar to those of the common-base configuration. The first step is to set up the direction of I_E as indicated by the arrow inside the *pnp* transistor symbol, as shown in Fig. 3.17 (a). Next, I_C and I_B are introduced to conform to Eq. (3.1), i.e., $I_E = I_C + I_B$, as shown in Fig. 3.17 (b). Finally, the voltages V_{BB} and V_{CC} are determined with polarities that will support the directions of I_B and I_C, as shown in Fig. 3.17 (c) to finish the biasing steps. If an *npn* transistor is involved, all the currents and polarities of Fig. 3.17 are reversed [1, 11].

The steps of biasing for the common-collector configuration are almost the same as those for the common-emitter configuration. Once the desired DC voltages and currents have been determined, a network must be set up that will provide the desirable

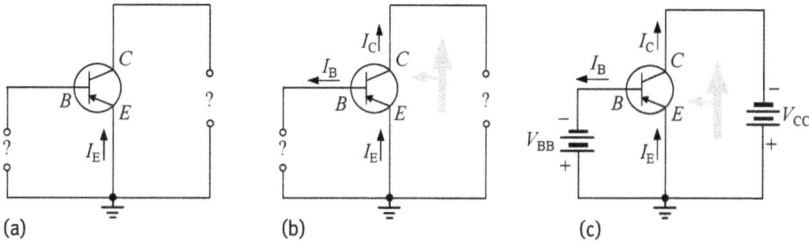

Fig. 3.17: Steps to determine biasing for common-emitter configuration, (a) Step 1, (b) Step 2, (c) Step 3.

operating point, that is, proper and practical (V_{CE}, I_C), based on which amplification is performed.

There are several widely-used practical biasing circuits and they will be discussed in the following sections.

3.3.3 Fixed-bias circuit

The simplest transistor DC biasing circuit, the fixed-bias circuit with a *pnp* transistor is illustrated in Fig. 3.18. Even though the configuration employs a *pnp* transistor, the calculations, equations and derivation apply equally well to an *npn* transistor network by simply changing voltage polarities and current directions.

The current directions in Fig. 3.18 are the actual current directions, and the voltage V_{CC} is negative, the same as the in DC biasing network of the common-emitter configuration shown in Fig. 3.10 (a).

The circuit in Fig. 3.18 is a practical circuit with AC input at the base terminal and output at the collector terminal. Due to their infinite DC reactance, the capacitors, C_1 and C_2, serve the purpose of isolation between DC biasing and AC amplification. In other words, the change of V_{CE}, I_C from the change of V_{CC}, R_B, or R_C, will be confined in the network between the two capacitors and will not influence the AC signals of v_{in}

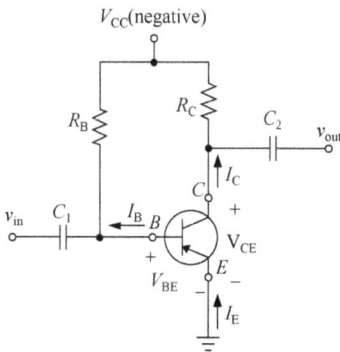

Fig. 3.18: Fixed-bias circuit with *pnp* transistor.

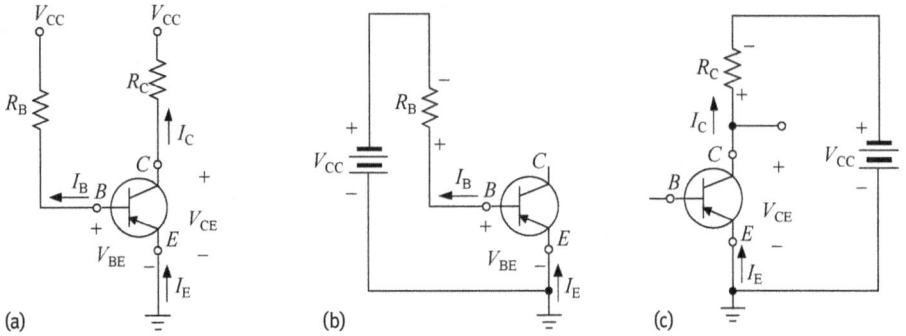

Fig. 3.19: DC analysis of a fixed-bias circuit, (a) DC equivalence, (b) Input loop, (c) Output loop.

or v_{out}. Moreover, for multiple-stage networks, the use of capacitors will isolate the Q-point setting within current stage, avoiding influence on neighboring stages.

For the DC analysis, the network can be extracted from the original network by replacing the capacitors, C_1 and C_2, with open circuits, as shown in Fig. 3.19 (a). In addition, the DC supply V_{CC} can be symbolically separated into two supplies, one connecting to R_B, the other to R_C, as shown in Fig. 3.19 (a). This way, input and output loops are separated but with the same power supply, V_{CC}. It also makes it easier to determine the base current I_B in the separated input loop.

First the input loop, i.e., the base-emitter circuit loop, is examined and redrawn as Fig. 3.19 (b). Applying Kirchhoff's voltage equation, it can be obtained that

$$V_{CC} = -I_B R_B + V_{BE}$$

Note that the emitter-base junction of the *pnp* transistor is forward biased, i.e., V_{BE} is negative and the absolute value is 0.7 V. The polarity of the voltage drop across R_B is established by the indicated direction of I_B. Solving the equation for the current I_B:

$$I_B = \frac{V_{BE} - V_{CC}}{R_B} \tag{3.8}$$

The physical meaning of Eq. (3.8) is clear: the numerator is the voltage across R_B, i.e., the applied voltage V_{CC} at one end less the drop across the base-to-emitter junction, V_{BE}; the denominator is the resistance of R_B. So, by Ohm's law, the base current is the voltage across R_B divided by the resistance of R_B. Note that $-V_{CC}$ in Eq. (3.8) is positive with larger absolute value and V_{BE} is negative with smaller absolute value, leading to a positive I_B.

Then, the collector-emitter loop is analyzed and redrawn as Fig. 3.19 (c) with the indicated direction of current I_C and the resulting polarity across R_C. Now, in the output loop, the magnitude of the collector current I_C is not obtained from the calculation. Instead, from Eq. (3.7), I_C is directly determined by I_B. It may not be easily accepted at first for newcomers to BJT, that the magnitude of I_C is not a function of the resistance

R_C, since the base current I_B is controlled by the level of R_B and I_C is related to I_B by a constant β, i.e., Eq. (3.7). Changing R_C will not affect the level of I_C or I_B so long as the conditions of the active region are satisfied. However, the level of R_C will definitely determine the magnitude of V_{CE}, which is a deterministic parameter to the operating point.

In the output loop as shown in Fig. 3.19 (c), applying Kirchhoff's voltage law, it is obtained that

$$V_{CC} = -I_C R_C + V_{CE}$$

So that

$$V_{CE} = V_{CC} + I_C R_C$$

which means that the voltage across the c-e junction is the supply voltage V_{CC} (negative) less the voltage drop across R_C (positive).

Now, I_C and V_{CE} are obtained and the Q-point, (I_C, V_{CE}), is set by a fixed-bias circuit. Although I_B is not directly used for setting the Q-point, it is a precondition to obtain I_C.

Moreover, since the base-emitter voltage drop V_{BE} and the supply voltage V_{CC} are fixed, the value of the base resistor R_B will determine the level of base current I_B, and then, the position of the operating point. This is also the reason for the name of this type of biasing circuit.

Example 3.1

For the fixed-bias network of Fig. 3.20, determine the following: I_{BQ}, I_{CQ}, V_{CEQ}, V_B, V_C and V_{BC}.

Solution 1

First, when the parameters, I_B, I_C and V_{CE}, are used specifically for the Q-point, they are referred to as I_{BQ}, I_{CQ} and V_{CEQ}.

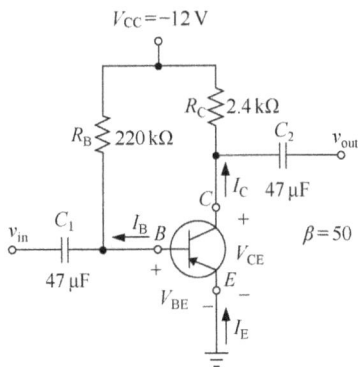

Fig. 3.20: Circuit of Example 3.1.

Then from the equations just derived, it is obtained that

$$I_B = \frac{V_{BE} - V_{CC}}{R_B} = \frac{-0.7\,V - (-12\,V)}{220\,k\Omega} = 51.36\,\mu A$$

and

$$I_C = \beta I_B = 50 \times 51.36\,\mu A = 2.568\,mA$$

and

$$V_{CE} = V_{CC} + I_C R_C = -12\,V + 2.568\,mA \times 2.4\,k\Omega = -5.84\,V$$

Because that emitter terminal is connected to ground, $V_E = 0\,V$. Also, $V_{BE} = -0.7\,V$. So,

$$V_B = V_{BE} + V_E = (-0.7\,V) + 0 = -0.7\,V$$

and

$$V_C = V_{CE} + V_E = (-5.84\,V) + 0 = -5.84\,V$$

and

$$V_{BC} = V_B - V_C = (-0.7\,V) - (-5.84\,V) = 5.14\,V$$

Now, from V_{BE} and V_{BC}, it is clear that the b-e junction is forward biased, and the c-b junction is reversed biased, as they should be for linear amplification.

Solution 2
From the analysis of Figs. 3.19 and 3.20, it can be seen that the network has established an equation that relates the variables I_C and V_{CE}; rearrange it here in the form of I_C as a function of V_{CE}:

$$I_C = \frac{V_{CE} - V_{CC}}{R_C} \tag{3.9}$$

On the other hand, BJTs possess output characteristics, as shown in Fig. 3.21, which nonlinearly relate the same two parameters of I_C and V_{CE}. Note that the output characteristics in Fig. 3.21, which are for the pnp transistor, have different orientations to those in Fig. 3.12, which are for the npn transistor. Therefore, the position of the Q-point simultaneously satisfies the network equation and the set of characteristics. From the viewpoint of mathematics, the Q-point is the solution of two simultaneous equations; one established by the network, Eq. (3.9) and the other by the BJT characteristics, Fig. 3.21.

Instead of solving the equations analytically, now a graphical way to do this will be introduced. The direct way is to find the intersection of the BJT output characteristics and the curve that corresponds to the network equation. This intersection satisfies both relationships and is the solution of the equation, i.e., the position of Q-point.

By examining Eq. (3.9), it can be seen that I_C is a linear function of V_{CE}, corresponding to a straight line. Moreover, it is known that a straight line is defined by

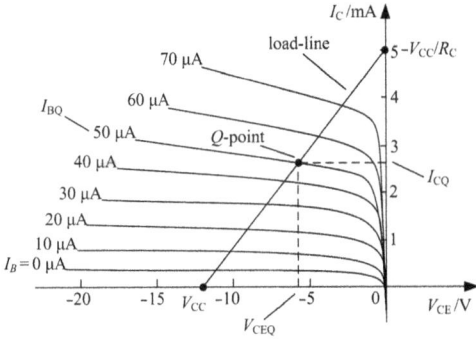

Fig. 3.21: Graphical solution to Example 3.1.

two points. So, finding two points from the equation, the straight line can be determined. For simplicity, letting variable and function be zero respectively, two points, $(0, I_C \mid V_{CE=0})$ and $(V_{CE} \mid I_{C=0}, 0)$ are obtained

$$I_C|_{V_{CE}=0} = \frac{V_{CE} - V_{CC}}{R_C} = \frac{(0) - V_{CC}}{R_C} = \frac{-(-12\,\text{V})}{2.4\,\text{k}\Omega} = 5\,\text{mA}$$

and

$$V_{CE}|_{I_C=0} = V_{CC} + I_C R_C = V_{CC} + (0) \cdot R_C = V_{CC} = -12\,\text{V}$$

Now, two points, (0 V, 5 mA) and (−12 V, 0 mA), are ready. A straight line can be drawn through them, as shown in Fig. 3.21.

Now, superimposing the straight line defined by the network on the characteristics, a series of intersections are obtained. Only the one corresponding to $I_{BQ} \approx 50\,\mu\text{A}$ (obtained by calculation in the same manner as solution 1) is the Q-point for the network. Then, from Q-point, drawing two dashed lines vertically and horizontally, respectively, two intersections are obtained on the abscissa and the ordinate, which result in parameters of the Q-point:

$$I_{CQ} \approx 2.6\,\text{mA}$$

$$V_{CEQ} \approx -5.8\,\text{V}$$

Moreover, V_B, V_C and V_{BC} can be calculated in the same manner as in solution 1.

It can be seen that the results from the graphical method are not exactly the same as those from the mathematical method. However, they are at the same level of accuracy. In later examples, the convenience of the graphical method will be made more obvious.

Moreover, as the straight line is determined by R_C, which is the load resistor of the common-emitter configuration, it is called the load line, and the graphical method is referred to as load-line analysis in other textbooks.

Generally speaking, by the graphical method, the influence on the position of the Q-point from circuit parameters, such as R_B, R_C and V_{CC}, can be seen more clearly.

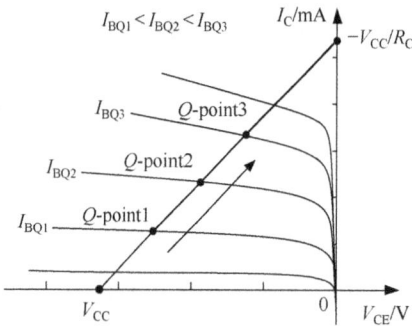

Fig. 3.22: Movement of the Q-point with different levels of I_B.

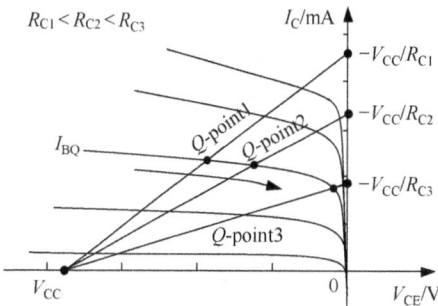

Fig. 3.23: Movement of the Q-point with different levels of R_C.

By varying the value of R_B, while keeping R_C and V_{CC} constant, the magnitude of I_B is changed, leading to the movement of the Q-point along the load line, as shown in Fig. 3.22. By changing R_C, while keeping R_B and V_{CC} constant, the load line will swing and make the Q-point move along the output curve, as shown in Fig. 3.23.

By changing V_{CC}, while keeping R_B and constant R_C, the load line will shift and make the Q-point move along the output curve, as shown in Fig. 3.24.

Now, another way of designing is very common in practical work, that is, find proper circuit parameters given a suitable Q-point.

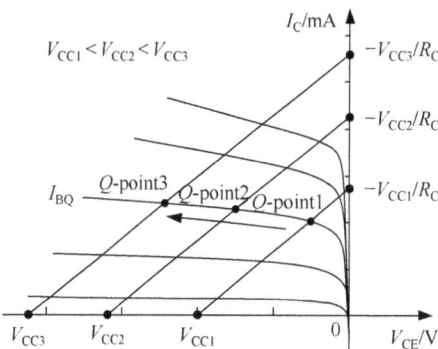

Fig. 3.24: Movement of the Q-point with different levels of V_{CC}.

Example 3.2

Shown in Fig. 3.25 is a suitable Q-point on the load line for the common-emitter configuration with the *pnp* transistor by a fixed-bias circuit. Find the required values of R_B, R_C and V_{CC}.

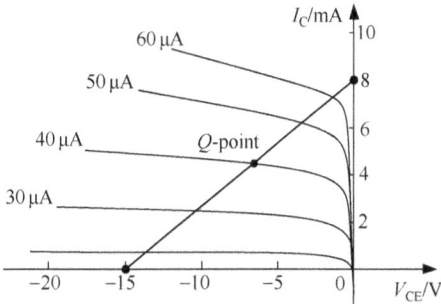

Fig. 3.25: Q-point of Example 3.2.

Solution

It is known that the intersection of the load line and abscissa should be V_{CC}. Also the intersection of load line and coordinate should be $-V_{CC}/R_C$. So, from Fig. 3.25, it can be obtained that

$$V_{CC} = -15\,V$$

and

$$\frac{-V_{CC}}{R_C} = 8\,mA$$

So,

$$R_C = \frac{(-V_{CC})}{8\,mA} = \frac{-(-15\,V)}{8\,mA} = 1.875\,k\Omega$$

Also, from Fig. 3.25,

$$I_{BQ} = 40\,\mu A$$

and

$$I_B = \frac{V_{BE} - V_{CC}}{R_B}$$

So,

$$R_B = \frac{V_{BE} - V_{CC}}{I_B} = \frac{-0.7\,V - (-15\,V)}{40\,\mu A} = \frac{-0.7\,V - (-15\,V)}{40\,\mu A} = 357.5\,k\Omega$$

However, the values of R_B and R_C obtained from the above calculation are not practical, since the values of resistors sold on the market are only several fixed standard values. So, for designing, the values of R_B and R_C should be selected as the nearest standard values to the calculated ones. Then, R_B should be selected as $1.8\,k\Omega$ and R_C as $360\,k\Omega$. Moreover, the selection of the resistor can keep the network within the range of accuracy with tolerance of errors. This is another reason why the graphical method is widely accepted in practical design, although it is not 100% accurate.

3.3.4 Emitter bias circuit

From the analysis of the fixed-bias circuit, the level of I_{BQ} is directly related to R_B. This means that the change of R_B for some reason will affect I_{BQ} and then the Q-point. So, this type of instability will harm the application of the network. Another type of DC bias network, the emitter bias circuit, shown in Fig. 3.26, contains one additional resistor connected to emitter terminal and can improve the stability level over that of fixed-bias configuration. The improvement of stability will be illustrated through the analysis in this section.

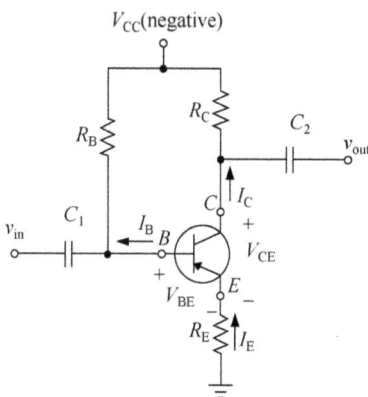

Fig. 3.26: Emitter bias circuit of the common-emitter configuration.

Even though the configuration employs a *pnp* transistor, the derivation, calculations and equations can equally apply to an *npn* transistor network by merely changing voltage polarities and current directions.

Note that the current directions in Fig. 3.26 are the actual current directions, and the voltage V_{CC} is negative.

To draw a DC equivalent circuit, capacitors, C_1 and C_2 are replaced with open circuits and DC supply V_{CC} is symbolically separated into two supplies, as shown in Fig. 3.27 (a). In this way, the base-emitter loop (input loop) and the collector-emitter loop (output loop) are separated, making it easier to perform DC analysis.

First, the input loop, i.e., the *b-e* loop, is examined and redrawn as Fig. 3.27 (b) with the indicated direction of current I_E and the resulting polarity across R_E. Applying Kirchhoff's voltage equation, we obtain

$$V_{CC} = -I_B R_B + V_{BE} - I_E R_E$$

From Eq. (3.1) and Eq. (3.7), we obtain

$$I_E = (\beta + 1)I_B \tag{3.10}$$

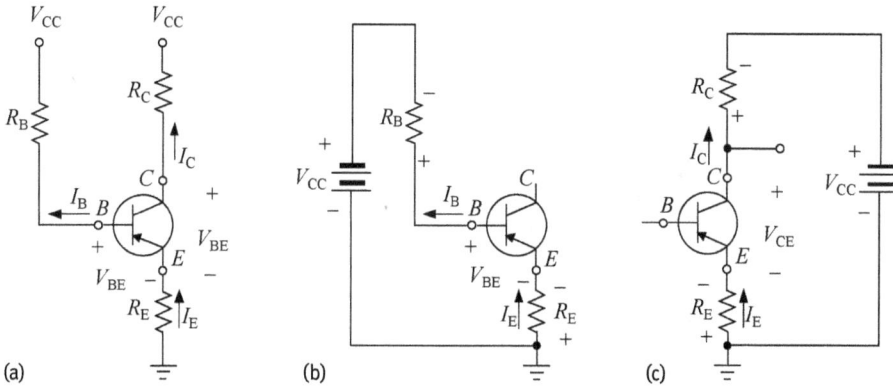

Fig. 3.27: DC analysis of the emitter bias circuit, (a) DC equivalence, (b) input loop, (c) output loop.

Substituting for I_E in the Kirchhoff's voltage equation results in

$$V_{CC} = -I_B R_B + V_{BE} - (\beta + 1)I_B R_E$$

So,

$$I_B = \frac{V_{BE} - V_{CC}}{R_B + (\beta + 1)R_E} \tag{3.11}$$

Note that V_{BE} is -0.7 V and the only difference between Eq. (3.11) for the emitter-bias configuration and Eq. (3.8) for the fixed-bias configuration is the term $(\beta + 1)R_E$.

Then, the *c-e* loop, redrawn as Fig. 3.27 (c), will be analyzed. Writing Kirchhoff's voltage law for the output loop results in

$$V_{CC} = -I_E R_E + V_{CE} - I_C R_C$$

So,

$$V_{CE} = V_{CC} + I_E R_E + I_C R_C$$

From Eq. (3.2), replacing I_E with I_C, we obtain

$$V_{CE} = V_{CC} + I_C(R_E + R_C) \tag{3.12}$$

which means that V_{CE} is the supply voltage V_{CC} (negative) less the voltage drop across R_C and R_E (both are positive).

As before, now I_C and V_{CE} are determined and the Q-point, (I_C, V_{CE}) is set by the emitter-bias circuit. This time, the relationship between I_C and I_E is involved.

Example 3.3

Shown in Fig. 3.28, it is the emitter bias network of the common-emitter configuration. Determine: I_B, I_C, V_{CE}, V_C, V_E, V_B and V_{BC}.

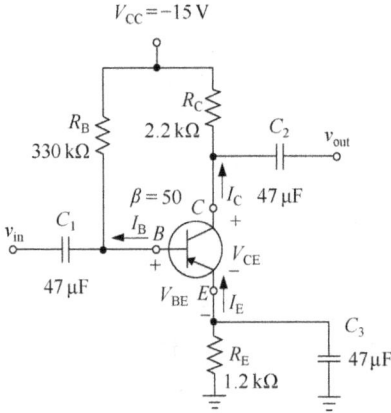

Fig. 3.28: Circuit of Example 3.3.

Solution 1

Note that in the network of Fig. 3.28, the capacitor C_3, connected in parallel with R_E, can be replaced with an open circuit in DC analysis. Its effect on AC analysis will be discussed in following sections. Figures 3.28 and 3.26 share the same DC equivalent circuit, as shown in Fig. 3.27.

Then, from the DC analysis of Fig. 3.27 and Eq. (3.11),

$$I_B = \frac{V_{BE} - V_{CC}}{R_B + (\beta + 1)R_E} = \frac{(-0.7 \text{ V}) - (-15 \text{ V})}{330 \text{ k}\Omega + (50 + 1)(1.2 \text{ k}\Omega)} = \frac{14.3 \text{ V}}{391.2 \text{ k}\Omega} = 36.55 \text{ μA}$$

So,

$$I_C = \beta I_B = 50 \times 36.55 \text{ μA} = 1.83 \text{ mA}$$

and

$$V_{CE} = V_{CC} + I_C(R_E + R_C) = -15 \text{ V} + 1.83 \text{ mA} \times (1.2 \text{ k}\Omega + 2.2 \text{ k}\Omega) = -8.78 \text{ V}$$

Then, from the top part of the loop in Fig. 3.27 (c),

$$V_C = V_{CC} + I_C R_C = (-15 \text{ V}) + 1.83 \text{ mA} \times 2.2 \text{ k}\Omega = -10.97 \text{ V}$$

and

$$V_E = V_C - V_{CE} = (-10.97 \text{ V}) - (-8.78 \text{ V}) = -2.19 \text{ V}$$

Or, from the lower part of the loop in Fig. 3.27 (c),

$$V_E = -I_E R_E \approx -I_C R_E = -1.83 \text{ mA} \times 1.2 \text{ k}\Omega = -2.196 \text{ V}$$

Both ways of calculating of V_E come to the same result. Moreover,

$$V_B = V_E + V_{BE} = (-2.2 \text{ V}) + (-0.7 \text{ V}) = -2.9 \text{ V}$$

Then,

$$V_{BC} = V_B - V_C = (-2.9 \text{ V}) - (-10.97 \text{ V}) = 8.07 \text{ V}$$

Now, for the *pnp* transistor, V_{BE} (negative) and V_{BC} (positive) are obtained, and it is clear that the *b-e* junction is forward biased, and the *c-b* junction is reversed biased, as they are required by linear amplification application of BJT.

Solution 2

Now the graphical method will be introduced. From the viewpoint of the network, Eq. (3.12) can be rearranged in the form of I_C as a function of V_{CE}:

$$I_C = \frac{V_{CE} - V_{CC}}{R_E + R_C}$$

So letting variable and function be zero, respectively, two points, $(0, I_C \mid V_{CE=0})$ and $(V_{CE} \mid I_{C=0}, 0)$ will be obtained:

$$I_C|_{V_{CE}=0} = \frac{V_{CE} - V_{CC}}{R_E + R_C} = \frac{0 - (-15\,V)}{1.2\,k\Omega + 2.2\,k\Omega} = 4.41\,mA$$

Moreover,

$$V_{CE}|_{I_C=0} = V_{CC} + I_C(R_E + R_C) = V_{CC} + (0) \cdot R_C = V_{CC} = -15\,V$$

Now, two points, $(0\,V, 4.41\,mA)$ and $(-15\,V, 0\,mA)$, are ready, and a straight line can be drawn through them, as shown in Fig. 3.29. Through the same steps in solution 1, $I_{BQ} \approx 37\,\mu A$ can be obtained. So, the Q-point is set. Then from the Q-point,

$$I_{CQ} \approx 1.8\,mA$$

$$V_{CEQ} \approx -8.8\,V$$

and the remaining V_B, V_C, V_E and V_{BC} can be calculated in the same manner as in solution 1.

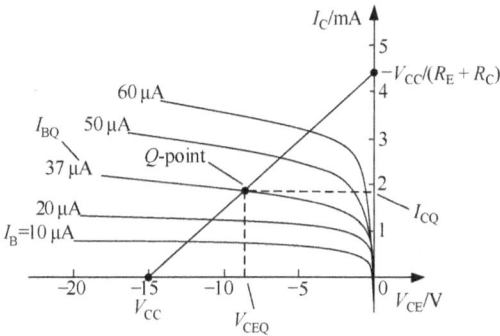

Fig. 3.29: Graphical solution to Example 3.3.

Moreover, back to the emitter-bias circuit, the additional resistor R_E connected to the emitter can improve the stability of DC biasing. The biasing voltages and currents will not move far away from the values set by the Q-point when parameter β changes due to temperature changes.

3.3.5 Voltage-divider bias circuit

In the previously discussed bias circuits, i.e., fixed-bias and emitter-bias circuits, the parameter β was used in the sequence of the calculation of the Q-point, that is, (I_{CQ}, V_{CEQ}) is a function of the parameter β. However, β is temperature sensitive, especially for silicon transistors, and the actual value of β for one specific transistor covers a range around the average value. The determination of the exact value of β is almost impossible. So, in order to set a relatively accurate and stable Q-point, it would be desirable to design a type of bias circuit that is less dependent on, or even independent of, β.

As shown in Fig. 3.30, the voltage-divider bias configuration is such a network that can set the Q-point with small or little dependency of β [1, 11].

To perform DC analysis of the voltage-divider configuration, there are two methods, one from the viewpoint of circuit theory, and the other a practical application. The former utilizes the knowledge of circuit analysis, such as the Thevenin equivalent circuit, to carry out an exact analysis. The conclusion is accurate, but the process is lengthy. The latter applies approximation to analysis, thus simplifying the process. The result obtained is at an acceptable level of accuracy, although not 100% correct, but with much less effort.

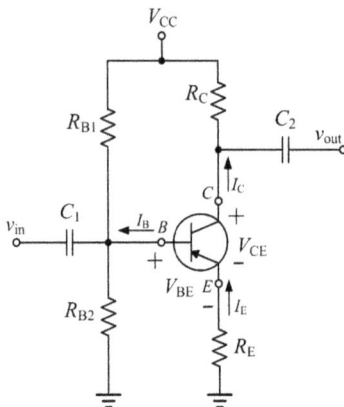

Fig. 3.30: Voltage-Divider bias circuit for common-emitter configuration.

Equivalent method

For the DC analysis, the input loop of the voltage-divider bias circuit can be redrawn as in Fig. 3.31 (a). According to Thevenin equivalent theory, the Thevenin equivalent network for the input loop can be set up as in Fig. 3.31 (b); it contains only two parameters, E_{TH} and R_{TH}. E_{TH} is the voltage of the Thevenin equivalent circuit, which is the voltage across the port when the right-hand side of the circuit is separated from the left-hand side, as shown in Fig. 3.32 (a). Actually, E_{TH} is just the voltage across R_{B2}

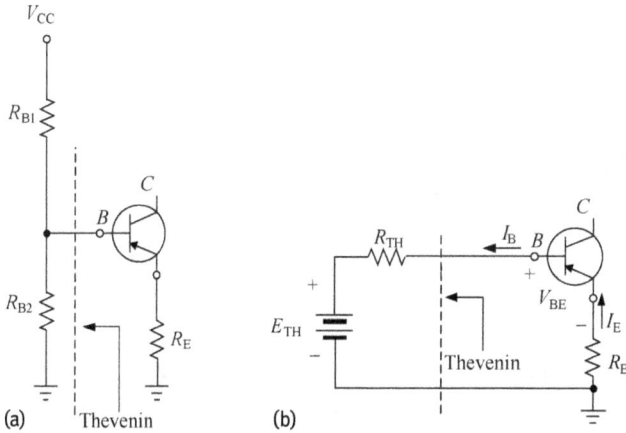

Fig. 3.31: Input loop of voltage-divider bias circuit, (a) DC equivalence, (b) Thevenin equivalent network.

from source V_{CC}. So,

$$E_{TH} = V_{RB2} = V_{CC}\frac{R_{B2}}{R_{B1} + R_{B2}} \tag{3.13}$$

and R_{TH} is the equivalent resistance from the port when the voltage source V_{CC} is set to ground, as shown in Fig. 3.32 (b). So,

$$R_{TH} = R_{B1} \parallel R_{B2} = \frac{R_{B1} \cdot R_{B2}}{R_{B1} + R_{B2}} \tag{3.14}$$

Going back to Fig. 3.31 (b), after the determination of Thevenin equivalent circuit parameters, E_{TH} and R_{TH}, I_{BQ} can be calculated by applying Kirchhoff's voltage law in the loop:

$$E_{TH} + I_B R_{TH} - V_{BE} + I_E R_E = 0$$

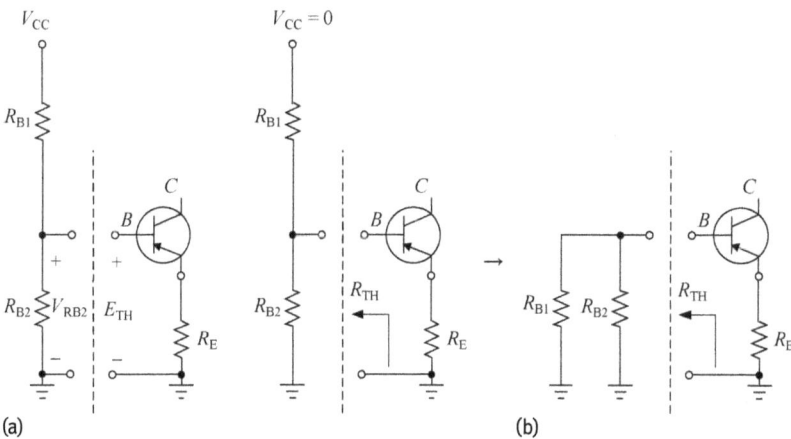

Fig. 3.32: Determination of Thevenin equivalent circuit parameters, (a) E_{TH}, (b) R_{TH}.

From Eq. (3.10), substituting I_E into the equation and rearranging in the form of I_B yields

$$I_B = \frac{V_{BE} - E_{TH}}{R_{TH} + (\beta + 1)R_E} \tag{3.15}$$

Although Eq. (3.15) initially seems to be different from Eq. (3.11) for the emitter-bias circuit, the numerator is still the difference between two voltages and the denominator is the base equivalent resistance plus the emitter resistor R_E by a factor of $(\beta+1)$, which looks like Eq. (3.11) obtained from the emitter-bias configuration.

Note that in Eq. (3.15), for the *pnp* transistor network, E_{TH} and V_{BE} are both negative.

Once I_B is known, I_C can be obtained from Eq. (3.7). Other parameters in the output loop can be calculated in the same way as developed for the emitter-bias configuration. For V_{CE}, Eq. (3.12) is still applicable, that is,

$$V_{CE} = V_{CC} + I_C(R_E + R_C)$$

The remaining equations for V_E, V_C and V_B can also be derived in the same way as for the emitter-bias configuration.

Approximation method
The input loop of the voltage-divider configuration can be redrawn as Fig. 3.33. Also, the currents, I_1, I_2 and I_B, are shown for the *pnp* transistor. The resistance R_{BE} is the equivalent resistance between base and ground with the emitter resistor R_E.

From the discussion of the emitter-bias circuit, especially Eq. (3.11), the equivalent resistance between base and emitter is defined as

$$R_{BE} = (\beta + 1)R_E$$

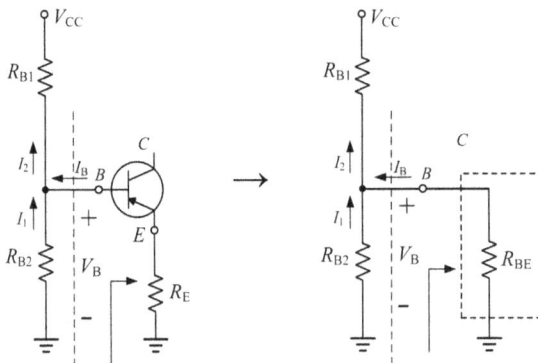

Fig. 3.33: Equivalent resistance between base and ground of the voltage-divider configuration.

If R_{BE} is much larger than the resistor R_{B2}, the current I_B will be much smaller than I_1. Then, I_2 will be approximately equal to I_1. So, the assumption is acceptable that I_B is essentially zero compared to I_1 or I_2. Then, $I_1 = I_2$, and R_{B1} and R_{B2} can be regarded as serially connected components. The voltage across R_{B_2}, which is actually the base voltage V_B, can be obtained using the voltage-divider rule,

$$V_B = V_{CC} \frac{R_{B2}}{R_{B1} + R_{B2}}$$

Now it is clear that the value of V_B is the same as E_{TH} derived in the equivalent method. Therefore, the most obvious difference between the equivalent and approximation methods is the separation effect from R_{TH} between E_{Th} and V_B when the Thevenin equivalent circuit is applied in the analysis.

Moreover, since the large ratio between R_{BE} and R_{B2} is the key to the approximation and $R_{BE} = (\beta + 1)R_E \approx \beta R_E$, the condition that should be satisfied to perform approximation is

$$\beta R_E \geq 10 R_{B2}$$

In other words, if β times R_E is at least 10 times larger than R_{B2}, the approximation method can be applied with a good level of accuracy.

Finally, once V_B is determined, V_E can be calculated from

$$V_E = V_B - V_{BE}$$

and the emitter current of I_E can be obtained

$$I_E = \frac{-V_E}{R_E}$$

Note that V_E is not the voltage drop across R_E, but $-V_E$ is. Also,

$$I_C \approx I_E$$

So, for V_{CE}, Eq. (3.12) is still applicable, that is,

$$V_{CE} = V_{CC} + I_C(R_E + R_C)$$

Note that in the sequence of calculations β does not appear and I_B was not calculated. Therefore, the determination of the Q-point is independent of the value of β, although it is involved in the judgment of the approximation condition.

Also, in the process of obtaining V_B, the voltage-divider rule of circuit analysis is involved. This is the reason for naming the bias circuit, the voltage-divider bias circuit.

Example 3.4

Determine the Q-point, that is, (I_{CQ}, V_{CEQ}) for the voltage-divider circuit of the common-emitter configuration shown in Fig. 3.34.

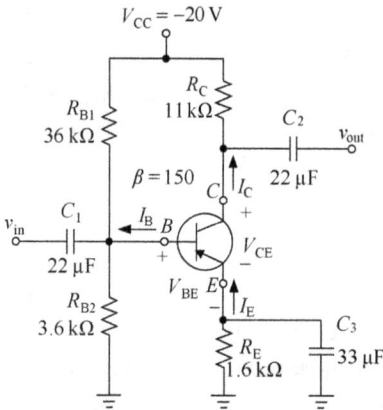

Fig. 3.34: Circuit of Example 3.4.

Solution 1

From the discussion of the equivalent method of the voltage divider and Eqs. (3.13) and (3.14) we obtain

$$E_{TH} = V_{CC} \frac{R_{B2}}{R_{B1} + R_{B2}} = (-20\,V) \cdot \frac{3.6\,k\Omega}{36\,k\Omega + 3.6\,k\Omega} = -1.82\,V$$

and

$$R_{TH} = \frac{R_{B1} \cdot R_{B2}}{R_{B1} + R_{B2}} = \frac{36\,k\Omega \cdot 3.6\,k\Omega}{36\,k\Omega + 3.6\,k\Omega} = 3.27\,k\Omega$$

Further, from Eq. (3.15) we obtain

$$I_B = \frac{V_{BE} - E_{TH}}{R_{TH} + (\beta + 1)R_E} = \frac{(-0.7\,V) - (-1.82\,V)}{3.27\,k\Omega + (150 + 1) \times 1.6\,k\Omega} = 4.57\,\mu A$$

Also, from Eq. (3.7),

$$I_C = \beta I_B = (150) \times (4.57\,\mu A) = 0.686\,mA$$

From Eq. (3.12),

$$V_{CE} = V_{CC} + I_C(R_E + R_C) = (-20\,V) + (0.686\,mA) \times (1.6\,k\Omega + 11\,k\Omega) = -11.36\,V$$

So the Q-point is at $(-11.36\,V, 0.686\,mA)$.

Solution 2

Before performing the approximation, the condition should be tested.

$$\beta R_E = 150 \times 1.6\,k\Omega = 240\,k\Omega$$

and,

$$10R_{B2} = 10 \times 3.6\,k\Omega = 36\,k\Omega$$

So the inequality,

$$\beta R_E \geq 10 R_{B2}$$

is satisfied. Then,

$$V_B = V_{CC} \frac{R_{B2}}{R_{B1} + R_{B2}} = (-20\,\text{V}) \cdot \frac{3.6\,\text{k}\Omega}{36\,\text{k}\Omega + 3.6\,\text{k}\Omega} = -1.82\,\text{V}$$

Then,

$$V_E = V_B - V_{BE} = (-1.82\,\text{V}) - (-0.7\,\text{V}) = -1.12\,\text{V}$$

and the emitter current of I_E can be obtained

$$I_E = \frac{-V_E}{R_E} = \frac{1.12\,\text{V}}{1.6\,\text{k}\Omega} = 0.7\,\text{mA}$$

and

$$I_C \approx I_E = 0.7\,\text{mA}$$

Then,

$$V_{CE} = V_{CC} + I_C(R_E + R_C) = (-20\,\text{V}) + (0.7\,\text{mA}) \times (1.6\,\text{k}\Omega + 11\,\text{k}\Omega) = -11.18\,\text{V}$$

So the Q-point is at $(-11.18\,\text{V}, 0.7\,\text{mA})$ by the approximation method.

Compare the Q-point by the equivalent method, $(-11.36\,\text{V}, 0.686\,\text{mA})$; they are certainly close. Moreover, considering the actual inaccuracy of parameter values, results from both methods can be trusted. The larger the ratio between βR_E and $10\,R_{B2}$, the closer the results from the approximation method are to those from the equivalent method.

Finally, from the previous discussions it is clear that the level of I_{BQ} will change with the change in β. However, the Q-point, defined by (I_{CQ}, V_{CEQ}), can remain almost constant if proper network parameters are selected to satisfy the approximation condition. Higher stability of the Q-point is the most attractive property of the voltage-divider biasing circuit.

3.3.6 Common-base configuration

Now, for simplicity, an example is used to show the analysis of the biasing circuit for common-base configuration. The results obtained can be applicable to other common-base configurations.

Example 3.5
Fig. 3.35 shows the common-base configuration. Determine I_B and V_{CB}.

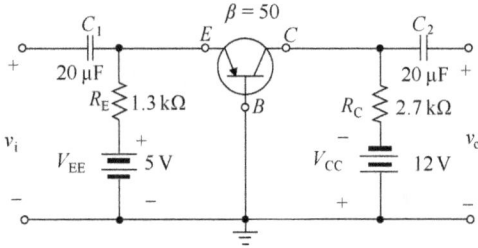

Fig. 3.35: Circuit of Example 3.5.

Solution

Before calculation, the DC equivalent circuits are as illustrated in Fig. 3.36.

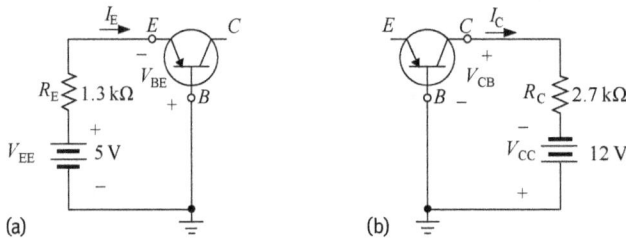

Fig. 3.36: DC equivalent circuit of Example 3.5, (a) Input loop, (b) Output loop.

For the input loop, as shown in Fig. 3.36 (a), by applying Kirchhoff's voltage law, we obtain

$$V_{EE} = I_E R_E - V_{BE}$$

So,

$$I_E = \frac{V_{EE} + V_{BE}}{R_E} = \frac{4\,V + (-0.7\,V)}{1.3\,k\Omega} = 2.54\,mA$$

Note that the polarity of V_{BE} is negative, due to the *pnp* transistor involved.

For the output loop, as shown in Fig. 3.36 (b), again applying Kirchhoff's voltage law, we obtain

$$V_{CC} = I_C R_C - V_{CB}$$

So,

$$V_{CB} = I_C R_C - V_{CC} \approx I_E R_C - V_{CC} = 2.54\,mA \times 2.7\,k\Omega - 12\,V = -5.14\,V$$

Note that polarity of V_{CB} is negative, due to the reverse-biased *c-b* junction of the *pnp* transistor.

From Eq. (3.7),

$$I_B = \frac{I_C}{\beta} \approx \frac{I_E}{\beta} = \frac{2.54\,mA}{50} = 50.8\,\mu A$$

3.3.7 Common-collector configuration

As illustrated in Fig. 3.14, the common-collector configuration can be drawn in a manner similar to the common-emitter configuration, with the load resistor connected between emitter and ground, the collector wired to AC ground, and the input signal fed from the base terminal. Now this configuration will be examined in detail for DC biasing.

Example 3.6
Figure 3.37 shows the common-collector configuration or emitter follower. Determine I_E and V_{CEQ}.

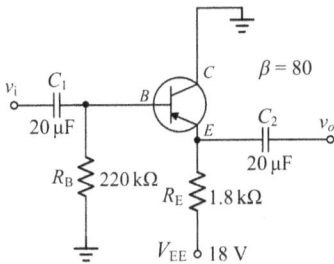

Fig. 3.37: Circuit of Example 3.6.

Solution
The DC equivalent circuits are illustrated in Fig. 3.38. For the input loop, as shown in Fig. 3.38 (a), by applying Kirchhoff's voltage law, we obtain

$$V_{EE} = I_E R_E - V_{BE} + I_B R_B$$

From Eq. (3.10)

$$I_E = (\beta + 1)I_B$$

So,

$$V_{EE} = (\beta + 1)I_B R_E - V_{BE} + I_B R_B$$

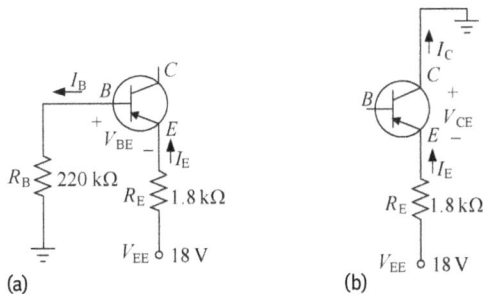

Fig. 3.38: DC equivalent circuit of Example 3.6, (a) Input loop, (b) Output loop.

Then,
$$I_B = \frac{V_{EE} + V_{BE}}{R_B + (\beta + 1)R_E} = \frac{18\,V + (-0.7\,V)}{220\,k\Omega + (80 + 1) \times 1.8\,k\Omega} = 47.29\,\mu A$$

Note that the polarity of V_{BE} is negative due to the *pnp* transistor involved. Also,

$$I_E = (\beta + 1)I_B = (80 + 1) \times 47.29\,\mu A = 3.83\,mA$$

Then in the output loop, as shown in Fig. 3.38 (b), again by applying Kirchhoff's voltage law, we obtain
$$V_{EE} = I_E R_E - V_{CE}$$

So,
$$V_{CE} = I_E R_E - V_{EE} = (\beta + 1)I_B R_E - V_{EE}$$
$$= (80 + 1) \times 47.29\,\mu A \times 1.8\,k\Omega - 18\,V$$
$$= -11.11\,V$$

Then, the results are
$$I_E = 3.83\,mA$$
$$V_{CEQ} = -11.11\,V$$

3.4 BJT AC analysis

3.4.1 Introduction to AC analysis

The fundamentals of the bipolar junction transistor were introduced in Section 3.1. The configurations and characteristics of BJT were examined in Section 3.2. In Section 3.3, various DC biasing circuits were discussed. Now it is time to examine the AC response of the BJT amplifier by applying the most frequently used AC equivalent circuits, or so-called models, to represent the transistor in the sinusoidal AC domain.

However, before introducing transistor AC models, it is more necessary to give a clear explanation of transistor amplification. Taking the common-base configuration of the *pnp* transistor shown in Fig. 3.39 as an example, the relationship between I_C and I_E is known and the input and output characteristics have been given.

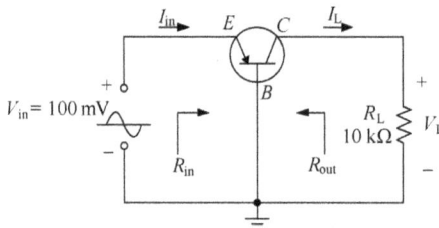

Fig. 3.39: Voltage amplifier of common-base configuration by *pnp* transistor.

Now the emphasis is the AC response of the network, so the DC biasing circuit can be ignored. In the input side, the AC input resistance determined by the characteristics of Fig. 3.7 is rather small, due to the almost vertical curves, normally lower than 100 Ω. In the output side, due the horizontal curves of the characteristics in Fig. 3.8, the AC output resistance is quite high, ranging from 50 kΩ to 1 MΩ. The difference between the input and output resistance can also be explained as the result of the different biasing conditions of the *b-e* and *c-b* junctions, that is, the former is forward biased and the latter is reversed biased [3, 11, 12].

To describe the amplifier mathematically, assuming the input resistance as 50 Ω,

$$I_{in} = \frac{V_{in}}{R_{in}} = \frac{100\,mV}{50\,\Omega} = 2\,mA$$

and from $I_{in} = I_E$, $I_E \approx I_C$ and $I_C = I_L$ we obtain

$$
\begin{aligned}
V_L &= I_L R_L \\
&= I_C R_L \\
&\approx I_E R_L \\
&= I_{in} R_L \\
&= 2\,mA \times 10\,k\Omega \\
&= 20\,V
\end{aligned}
$$

So, the voltage amplification ratio is

$$
\begin{aligned}
A_v &= \frac{V_L}{V_{in}} \\
&= \frac{20\,V}{100\,mV} \\
&= 200
\end{aligned}
$$

This means that the output voltage is 200 times larger than the input voltage. The typical voltage amplification ratio of the common-base configuration ranges between 50 to 300. However, the current amplification ratio (I_L/I_{in}) of the common-base configuration is normally less than unity, since $I_L/I_{in} = I_C/I_E = \alpha$.

From the explanation of the amplification process, it can be seen that the input and output resistance greatly influence the input and output voltages in the case when the input and output currents are similar. So, the basic amplifying function by the circuit is acquired by transferring currents from a smaller resistor to a larger one. So, the core device of the circuit that can "transfer a resistor" is referred to as the "transistor", of which the basic function is amplification.

Moreover, in practical applications, AC voltages and currents are more relevant, instead of DC ones. Also, the magnitude of the input sinusoidal signal has the first priority, which affects the types of circuit. Normally, there are two types of circuits involving AC signals, small-signal and large-signal ones. The magnitude of the input signal is compared with the scales of the device characteristics to judge whether it belongs to small or large signals. Sometimes the dividing line between the two is not very clear. However, the application will usually make it quite clear which technique is more appropriate. The small-signal techniques involve amplification of weak input signals, which are normally outputs of various sensors. After amplification to sufficiently high magnitudes, they will be processed more conveniently in subsequent stages; while large-signal techniques specifically provide relatively smaller input signals with sufficient power, which are used to drive high-power peripherals [3, 12].

Another issue related to transistor amplifiers is where the increased amount of power in the output signal comes from. As stated in the law of conservation of energy, that energy can neither be created nor destroyed; rather, it transforms from one form to another. Actually, the increased amount of power in the output signal comes from the DC supply. It is the transistor that converts the energy from the DC power supply to the AC output signal [3].

Actually, DC and AC electrical levels exist simultaneously in the circuit. So, the superposition theorem is applicable for the analysis of both DC and AC components of the BJT network, making it possible to separate the analysis of the DC and AC responses of the network. A realistic way is that one can make a complete DC analysis of a network, such as setting the Q-point, before considering the AC analysis, such as calculating amplification ratio. However, circuit parameters such as the output resistance appearing in the AC analysis will actually be affected by the DC conditions. This means that there is still a tight link between DC and AC analysis [11, 12].

The discussion in the previous sections of this chapter is mainly on DC analysis. The AC part, which belongs to the small-signal technique, will follow subsequently. Large-signal analysis, or power amplification, will not be included in this textbook.

3.4.2 AC equivalent circuits

The central part to small-signal AC analysis of transistors is the introduction of equivalent circuits or circuit models. A circuit model is defined as a specifically designed network of basic circuit elements, which best describes or approximates the actual behavior of a semiconductor device under specific operation conditions and facilitates the analysis of the device by avoiding some trivial details.

The way to use the model is simple and straightforward. Once the AC equivalent circuit of the network has been determined, the electrical symbol for the device can be replaced by its model. Then basic knowledge of circuit analysis can be applied to find the desired parameters of the network.

There are three widely-used models in the small-signal AC analysis of semiconductor transistors: the r_e model, the hybrid π model and the hybrid equivalent model. Throughout the text only the r_e model will be emphasized. Also note that once profi-

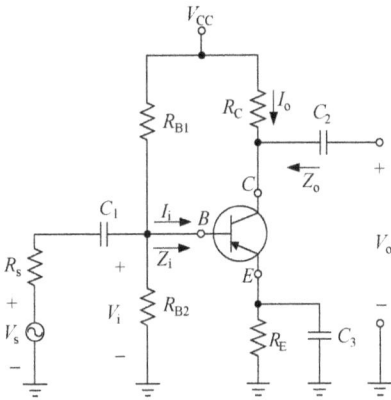

Fig. 3.40: Circuit of which the AC response is to be examined.

ciency with one model has been gained, it will not be a dramatic undertaking to carry over to the investigation of a different model.

Before introducing small-signal AC equivalent circuits for transistors, it is necessary to show how to change the original circuit, shown in Fig. 3.40, to its AC equivalent circuit. Also, small-signal AC parameters such as V_i, V_o, I_i, I_o, Z_i and Z_o, are shown in Fig. 3.40 and are used to evaluate the performance of the AC response. The input voltage, V_i, is defined as the voltage between base and ground; the input current, I_i, as the base current of the transistor; the input impedance, Z_i, as from base to ground; the output voltage, V_o, as the voltage between collector and ground; the output current, I_o, as the current through the load resistor R_C; and the output impedance, Z_o, as from collector to ground.

First, change the DC supplies to ground. The DC levels are only indispensable for determining a proper Q-point. After the determination of the Q-point, the DC levels can be ignored in the AC equivalent circuit. Also, the DC levels in output voltages, resulting from the DC levels of the input voltages, are of little significance from the viewpoint of information. So, all the DC supplies can be replaced by zero-voltage potentials or ground, as illustrated in Fig. 3.41 (a).

In addition, the coupling capacitors C_1 and C_2 and the bypass capacitor C_3 are assumed to have a very small reactance at the frequency of application. So, they can be replaced by a short circuit for all practical purposes. The replacement of C_3 will result in the "shorting out" of the DC biasing resistor R_E in the AC equivalent circuit only, as shown in Fig. 3.41 (b). Note that the obvious different equivalent circuits for capacitors in DC and AC response are the "open-circuit" equivalent under DC steady-state conditions, resulting in isolation between stages for the quiescent conditions and DC levels, and the "short-circuit" equivalent for AC response, resulting in a smooth path for AC signals.

Although the network appearance may change, such AC quantities as V_i, V_o, I_i, I_o, Z_i and Z_o, should be kept the same as defined by the original network. Then to simplify

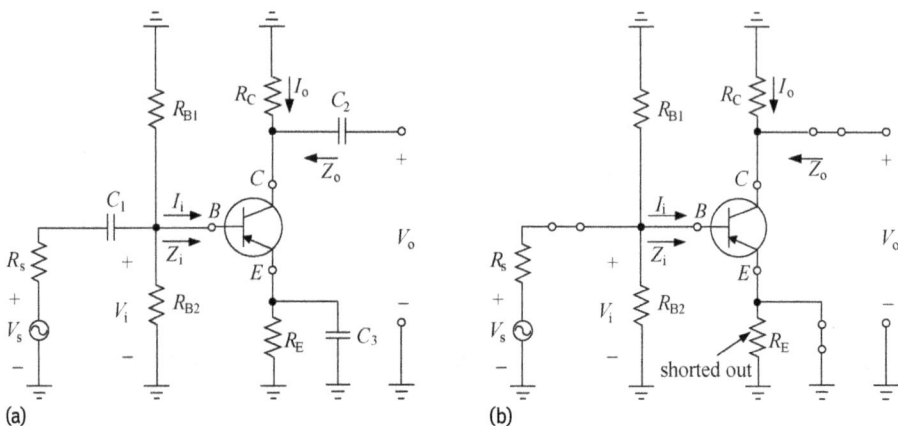

(a)   (b)

Fig. 3.41: Examination of an AC equivalent circuit, (a) Removing DC supplies, (b) Removing capacitors.

the AC equivalent circuit further, as shown in Fig. 3.42, a common ground can be used; R_{B1} and R_{B2} are rearranged as in parallel, and R_C will appear from collector to emitter.

Also shown in Fig. 3.42, the box shown by dashed lines represents the transistor equivalent circuit, or model, the components of which employ familiar components such as resistors and independent or controlled sources. So, analysis techniques such as superposition, Thevenin's theorem and so on, can be applied to find the desired parameters.

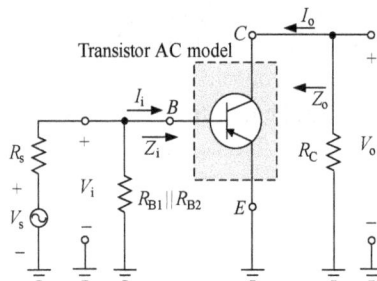

Fig. 3.42: Small-signal AC analysis of circuit in Fig. 3.40.

Since the transistor serves as an amplifying device, the quantity of voltage gain, $A_v = V_o/V_i$, is of more concern, because it indicate how large the output voltage V_o is compared with the input voltage V_i. Similarly, for this configuration, the current gain $A_i = I_o/I_i$, and $I_i = -I_b$, and $I_o = -I_c$. The input impedance Z_i and the output impedance Z_o are of particular importance in the analysis to follow. More details relating to these quantities will be presented in the following sections.

In short, the AC equivalent circuit of a network can be obtained through the following steps:

(1) Set all DC sources to zero and replace them by the short-circuit equivalent.
(2) Replace all capacitors by the short-circuit equivalent.
(3) Remove all elements bypassed by the short-circuit equivalents generated by steps 1 and 2.
(4) Rearrange the network in a more logical, concise and convenient form.

In the sections to follow, the transistor r_e model will be introduced to perform the AC analysis of the network of Fig. 3.40.

3.4.3 Transistor r_e model

Now the r_e model for the common-base, common-emitter and common-collector configurations of the BJT transistor will be introduced. The description of them is given in detail to approximate well enough the actual behavior of the BJT transistor. Note that in Fig. 3.42, the box shown with dashed lines has three sides that have has a BJT terminal through them. Also, the BJT model has three terminals that correspond to the three BJT terminals.

1. r_e model for common-base configuration
Shown in Fig. 3.6 is the common-base configuration for both types of BJT transistor. To concentrate on the BJT itself, the simplified common-base configuration for the *npn* transistor is shown in Fig. 3.43 (a) with six AC circuit parameters on input and output ports. It is worth mentioning that the base terminal has been extended as two terminals, one for the input port and the other one for the output port. Both are at the same electric potential.

After the introduction of the r_e model, it should be fairly obvious that isolation has been set up between the input emitter side and the output collector side. The interaction between the input and output portions is represented by a controlled current source with the relationship

$$I_c = \alpha I_e$$

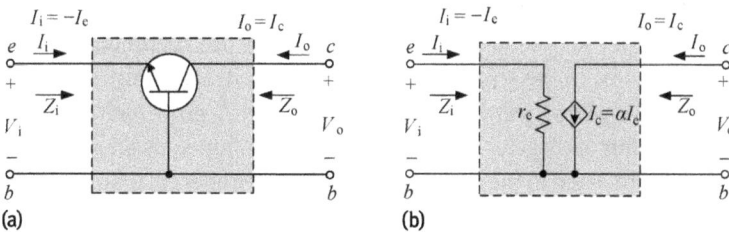

Fig. 3.43: r_e model for common-base configuration, (a) Simplified common-base circuit, (b) r_e model.

In this case, the collector current I_c is "controlled" by the level of emitter current I_e. Note that the diamond-shaped representation of the current source means that it is the controlled source, whereas the independent current source employs the circular enclosure.

On the other hand, on the input side of the model, an equivalent AC resistance r_e between emitter and base terminals has been introduced, the value of which can be found by the equation

$$r_e = \frac{26\,\text{mV}}{I_E} \tag{3.16}$$

indicating that it is the DC level of the emitter current I_E that determines the AC parameter of the resistance r_e. Also note that DC quantities are represented with subscripts of upper-case letters, such as I_C, I_E and V_{BE}, whereas AC quantities are represented by lower-case letters, such as r_e, I_i and I_o.

Due to the isolation that exists between input and output circuits, as shown in Fig. 3.43 (b), it is clear that the input impedance Z_i for the common-base configuration of the BJT is simply r_e, that is,

$$Z_i = r_e \tag{3.17}$$

Typical values of Z_i for the common-base configuration cover from a few ohm to a maximum of about 50 Ω.

To determine the output impedance Z_o, set the input signal to zero, that is,

$$I_e = 0\,\text{A}$$

So,

$$I_c = \alpha I_e = \alpha \cdot 0 = 0$$

leading to an open-circuit equivalence at the output terminals,

$$Z_o = \infty\,\Omega \tag{3.18}$$

In fact, typical values of Z_o for the common-base configuration are in the level of megaohm.

Moreover, from the viewpoint of the output characteristics of the common-base configuration shown in Fig. 3.8, the output impedance Z_o can be determined by the slope of characteristic lines. If the lines are assumed as horizontal, the result is just the same as in Eq. (3.18). If the lines are measured graphically or experimentally, the output impedance Z_o can be in the level of megaohm. Generally speaking, the input impedance Z_i of is relatively small, and the output impedance Z_o quite high.

Furthermore, the common-base configuration will be examined as an amplifier, with a load resistor R_L connected to the output port, to show the effect of amplification (Fig. 3.44).

First, the voltage gain A_v will be determined. In the output port,

$$V_o = -I_o R_L = -(I_c)R_L = -\alpha I_e R_L$$

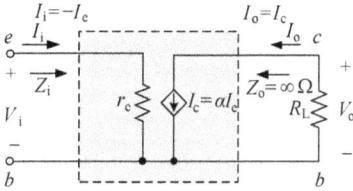

Fig. 3.44: Common-base configuration with a load resistor R_L.

In the input port,

$$V_i = I_iZ_i = (-I_e)Z_i = -I_e r_e$$

So, the voltage gain A_v for the common-base configuration is

$$A_v = \frac{V_o}{V_i} = \frac{-\alpha I_e R_L}{-I e r_e} = \alpha\frac{R_L}{r_e} \approx \frac{R_L}{r_e}$$

The positive polarity of A_v reveals that V_o and V_i are in phase for the common-base configuration. Then, for the current gain A_i,

$$A_i = \frac{I_o}{I_i} = \frac{I_c}{-I_e} = \frac{\alpha I_e}{-I_e} = -\alpha \approx -1$$

To give a clearer description of the property of the common-base amplifier, a *pnp* transistor is involved as shown in Fig. 3.45, with six AC circuit parameters on both input and output ports. Note that the main difference between Figs. 3.45 and 3.43 (a) is the directions of currents I_c and I_e, which may be the same or opposite to those of input and output currents I_i and I_o.

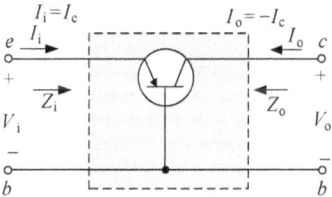

Fig. 3.45: Common-base amplifier with the *pnp* transistor.

2. r_e model for the common-emitter configuration

For the common-emitter configuration with the *pnp* transistor shown in Fig. 3.46 (a), the input signal is applied between the base and emitter terminals, whereas the output is set between the collector and emitter terminals. As indicated by its name, the emitter terminal is common to both the input and output ports of the amplifier.

Simply substituting the r_e equivalent circuit of the common-base amplifier results in the configuration of Fig. 3.46 (b). The controlled-current source is still connected between the collector and base terminals, with the relationship.

$$I_c = \beta I_b$$

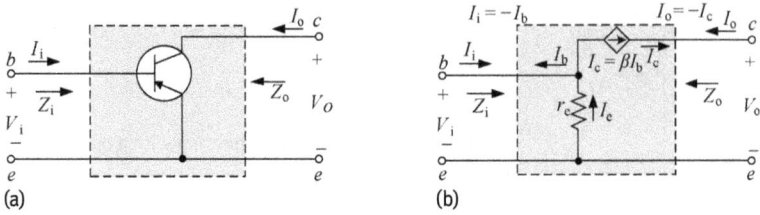

Fig. 3.46: Common-emitter amplifier with the *pnp* transistor, (a) AC parameters, (b) Approximate AC model.

The resistor r_e is connected between the base and emitter terminals. In this configuration, the base current I_b and the collector current I_c are opposite to the input current I_i and the output current I_o, respectively.

The input impedance Z_i can be calculated by

$$Z_i = \frac{V_i}{I_i}$$
$$= \frac{V_{be}}{-I_b}$$
$$= \frac{-I_e r_e}{-I_b}$$
$$= \frac{I_e r_e}{I_b}$$

From Eq. (3.10), substituting I_e into the equation, we have

$$Z_i = \frac{(\beta + 1)I_b r_e}{I_b}$$
$$= (\beta + 1)r_e$$

Also, β is usually sufficiently large and then the approximation, $\beta + 1 \approx \beta$, is acceptable. So that

$$Z_i \approx \beta r_e$$

It can be seen that the input impedance Z_i is β times the value of r_e. In other words, the resistance in the emitter terminal is reflected into the input loop by a multiplying factor of β. Typically for the common-emitter configuration, the Z_i value ranges from a few hundred ohm to the kiloohm range, with maxima less than $10\,k\Omega$.

To determine the output impedance Z_o, the output characteristics for the common-emitter configuration, as shown in Fig. 3.12, will be used. The slopes of the curves increase with an increase in the collector current I_C. The steeper (or higher) the slope, the lower (or smaller) is the level of output impedance Z_o. Typically for the common-emitter configuration, Z_o values are in the range 40–$50\,k\Omega$. If the output impedance Z_o is obtained graphically or from data sheets, it can be included as r_o, as illustrated in Fig. 3.47.

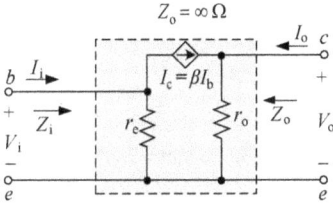

Fig. 3.47: Approximate AC model of the common-emitter configuration with r_o.

So in this case, when the input signal is set to zero and the current I_C is 0 A,

$$Z_o = r_o$$

However, if the role played by r_o is ignored as in Fig. 3.46 (b), we obtain

$$Z_o = \infty \, \Omega$$

Moreover, with a load resistor R_L connected to the output port, the common-emitter configuration will be examined as an amplifier, as shown in Fig. 3.48.

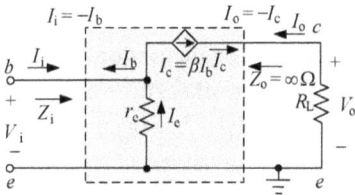

Fig. 3.48: Common-emitter amplifier with a load resistor R_L.

In the output port,

$$V_o = -I_o R_L = I_c R_L = \alpha I_e R_L$$

and in the input port, from Ohm's law, we obtain

$$V_i = (-I_e) r_e$$

So, the voltage gain A_v for the common-emitter configuration is

$$A_v = \frac{V_o}{V_i} = \frac{\alpha I_e R_L}{-I_e r_e} = -\alpha \frac{R_L}{r_e} \approx -\frac{R_L}{r_e}$$

The negative polarity of A_v reveals that V_o and V_i are 180° out of phase for the common-emitter configuration. Then, for the current gain A_i,

$$A_i = \frac{I_o}{I_i} = \frac{-I_c}{-I_b} = \frac{\beta I_b}{I_b} = \beta$$

Furthermore, to give a clearer conclusion of r_e model of the common-emitter configuration, including the parameters just discussed, the input impedance βr_e, the collector current βI_b and the output impedance r_o, Fig. 3.49 is given as the r_e model of the common-emitter configuration.

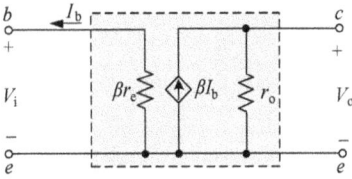

Fig. 3.49: r_e model of the common-emitter configuration.

Typically, the common-emitter configuration can be regarded as an amplifier with a moderate level of input impedance Z_i, and a high level of both voltage gain A_v and current gain A_i. Also, output impedance r_o may need to be included to facilitate the network analysis.

3. r_e model for common-collector configuration

For the common-collector configuration, the appropriate AC model defined for the common-emitter configuration in Fig. 3.46 can normally be applied, instead of defining a new model specifically for the common-collector configuration. The obvious reason is that they have a similar network structure. The following AC analysis will prove its effectiveness for both types of configurations.

After the introduction of the r_e models for the three BJT configurations, small-signal AC analysis will be performed on some standard BJT networks. These networks cover the majority of commonly used BJTs. Also, once the analysis has been well understood, modifications of the standard BJT network will be relatively easy to examine. In the following section, DC biasing circuits are used in combination with configurations as the names of the networks involved.

3.4.4 CE Configuration with fixed bias

The first configuration to be analyzed in detail is the common-emitter fixed-bias network of Fig. 3.50, which is the same as that in Fig. 3.18. However, now the discussion is from the viewpoint of AC analysis.

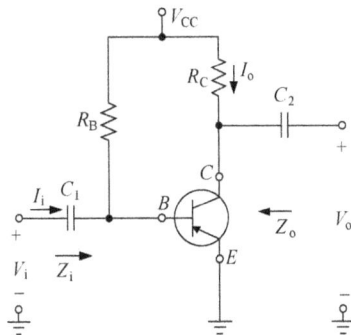

Fig. 3.50: Common-emitter fixed-bias configuration with AC parameters.

Now, for small-signal AC analysis, parameters are added to show the network properties: the input AC signal V_i is applied to the base terminal of the transistor, whereas the AC output V_o is off the collector. More importantly, note that due to the influence of R_B the input current I_i is not the base current, but the AC source current; and the output current I_o is opposite to the collector current I_c.

Now modify the original network to its AC equivalent circuit: change the DC supply V_{CC} to ground, replace the DC blocking capacitors C_1 and C_2 by the short-circuit equivalents. The AC equivalent circuit is shown in Fig. 3.51.

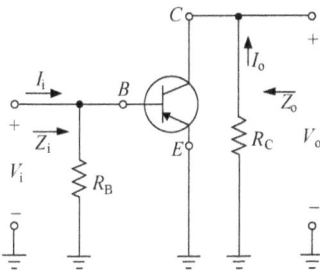

Fig. 3.51: AC equivalent circuit of common-emitter fixed-bias configuration.

In Fig. 3.51, it is obvious that the DC supply and the BJT emitter terminal share common ground, resulting in the relocation of R_B and R_C in parallel with the input and output ports of the network, respectively. In addition, note the positions of the AC parameters of Z_i, Z_o, I_i and I_o on the rearranged network, which should be the same as in original network of Fig. 3.50.

Now substituting the r_e model of Fig. 3.49 for the BJT of Fig. 3.51 results in the network of Fig. 3.52.

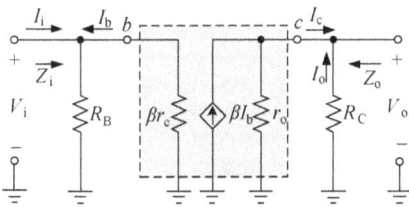

Fig. 3.52: Substitution of the r_e model into the network of Fig. 3.51.

Note that such parameters as β, r_e and r_o, are determined in the following way. The value of β is normally read from a specification sheet (or data sheet) or by measurement using a transistor testing instrument directly. The magnitude of r_e must be found from the DC analysis of the system, that is, Eq. (3.16). The value of r_o can normally be read from the data sheet or calculated from the characteristics. Now, assuming that β, r_e and r_o have been obtained, other AC parameters of the network can be determined.

Fig. 3.52 clearly shows that

$$Z_i = R_B \parallel \beta r_e$$

In most cases, R_B is greater than βr_e by at least a factor of 10. So, the following approximation is acceptable:

$$Z_i \approx \beta r_e$$

From the knowledge of circuit analysis, the output impedance of a network, Z_o, is defined as the equivalent impedance in the condition when all the voltage sources are set as short circuits and current sources as open-circuits. So, in Fig. 3.52, when the controlled current source is replaced with the open-circuit equivalence, Z_o can be obtained as

$$Z_o = R_C \parallel r_o$$

In the case of $r_o \geq 10\,R_C$, the following approximation is acceptable:

$$Z_o \approx R_C$$

In the output port, from Ohm's law, we obtain

$$V_o = (\beta I_b)(R_C \parallel r_o)$$

and in the input port,

$$V_i = (-I_b)\beta r_e$$

So,

$$V_o = \left(\beta \frac{-V_i}{\beta r_e} \right)(R_C \parallel r_o)$$

Then the voltage gain A_v is

$$A_v = \frac{V_o}{V_i} = \frac{-(R_C \parallel r_o)}{r_e} \tag{3.19}$$

In the case $r_o \geq 10\,R_C$, the following approximation is acceptable:

$$A_v = -\frac{R_C}{r_e} \tag{3.20}$$

Although β must be involved to determine r_e in the DC analysis, it is absent in the calculation of A_v.

For the common-emitter configuration, the negative polarity of A_v reveals that V_o and V_i are 180° out of phase. This is demonstrated in Fig. 3.53. Note that the amplitudes of V_o and V_i are opposite but with the same period of T, indicating that it is a linear amplification.

Example 3.7

For the common-emitter fixed-bias configuration, as shown in Fig. 3.54, determine the following parameters: r_e, Z_i, Z_o and A_v with $r_o = 50\,\text{k}\Omega$.

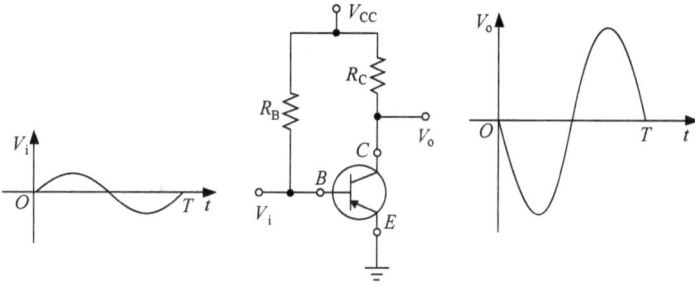

Fig. 3.53: 180° phase shift between waveforms of V_o and V_i.

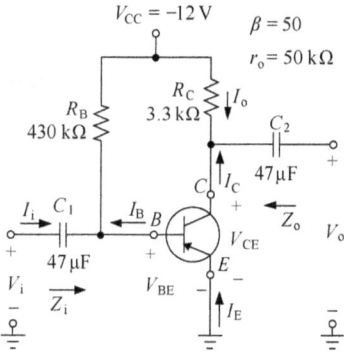

Fig. 3.54: Example 3.7.

Solution

First, some AC parameters are obtained from the DC analysis.

From Eq. (3.8),

$$I_B = \frac{V_{BE} - V_{CC}}{R_B} = \frac{(-0.7\,\text{V}) - (-12\,\text{V})}{430\,\text{k}\Omega} = 26.3\,\mu\text{A}$$

Then,

$$I_E = (\beta + 1)I_B = (100 + 1) \times 26.3\,\mu\text{A} = 2.65\,\text{mA}$$

So, from Eq. (3.16),

$$r_e = \frac{26\,\text{mV}}{I_E} = \frac{26\,\text{mV}}{2.65\,\text{mA}} = 9.81\,\Omega$$

Then, AC analysis is performed.

For Z_i,

$$Z_i = R_B \parallel \beta r_e = (430\,\text{k}\Omega) \parallel (100 \times 9.81\,\Omega) = 0.979\,\text{k}\Omega$$

Or to use the approximation, verify the condition:

$$\beta r_e = 100 \times 9.81\,\Omega = 0.981\,\text{k}\Omega$$

and

$$R_B = 430\,\text{k}\Omega$$

So,

$$R_B = 430\,k\Omega \geq 9.8\,k\Omega = 10 \times 0.98\,k\Omega = 10 \cdot \beta r_e$$

The condition is satisfied, so,

$$Z_i \approx \beta r_e = 0.98\,k\Omega$$

For Z_o,

$$Z_o = R_C \parallel r_o = (3.3\,k\Omega) \parallel (50\,k\Omega) = 3.10\,k\Omega$$

For A_v,

$$A_v = \frac{-(R_C \parallel r_o)}{r_e} = -\frac{(3.3\,k\Omega \parallel 50\,k\Omega)}{9.81\,\Omega} = -316$$

From the A_v, it can be seen that the output signal has been amplified, but out of phase with the input signal.

3.4.5 CE configuration with voltage-divider bias

Now from the viewpoint of the AC analysis, voltage-divider bias with common-emitter configuration will be investigated, as shown in Fig. 3.55.

Modify the original network in Fig. 3.55 to its AC equivalent circuit: change the DC supply V_{CC} to ground, replace the DC blocking capacitors C_1, C_2 and bypass capacitor C_E by the short-circuit equivalents. The AC equivalent circuit is shown in Fig. 3.56. Note that the resistor R_E does not exist in the AC equivalent circuit, because in the operation frequency range, the reactance of the capacitor C_E is so small compared to R_E that it can be replaced with a short-circuit across R_E. Moreover, R_{B1} and R_{B2} remain part of the input circuit, while R_C is part of the output circuit.

Substituting the r_e equivalent circuit leads to the network of Fig. 3.57.

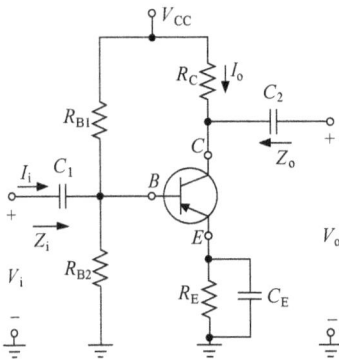

Fig. 3.55: Voltage-divider bias with common-emitter configuration with AC parameters.

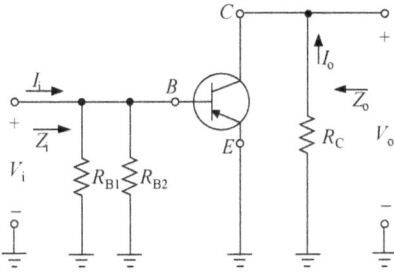

Fig. 3.56: AC equivalent circuit of voltage-divider bias with common-emitter configuration.

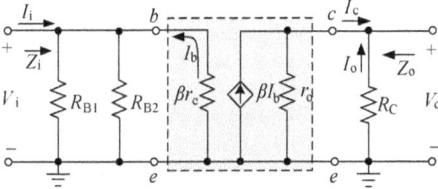

Fig. 3.57: Substitution of the r_e model into voltage-divider bias with common-emitter configuration.

Then, the AC parameters can be found. For Z_i, it is obvious that

$$Z_i = R_{B1} \parallel R_{B2} \parallel \beta r_e$$

For Z_o,

$$Z_o = R_C \parallel r_o$$

In the output port, with Ohm's law we obtain

$$V_o = (\beta I_b)(R_C \parallel r_o)$$

and in the input port,

$$V_i = (-I_b)\beta r_e$$

So that

$$V_o = \left(\beta \frac{-V_i}{\beta r_e}\right)(R_C \parallel r_o)$$

Then the voltage gain A_v is

$$A_v = \frac{V_o}{V_i} = \frac{-(R_C \parallel r_o)}{r_e} \tag{3.21}$$

Note that the AC parameters for the common-emitter fixed-bias network and the common-emitter voltage-divider bias network are almost the same, except for Z_i. The reason is obvious: the main difference between them exists only in the input side.

Example 3.8

For the common-emitter voltage-divider bias network, shown in Fig. 3.58, determine the following parameters: r_e, Z_i, Z_o and A_v with $r_o = 40\,\text{k}\Omega$.

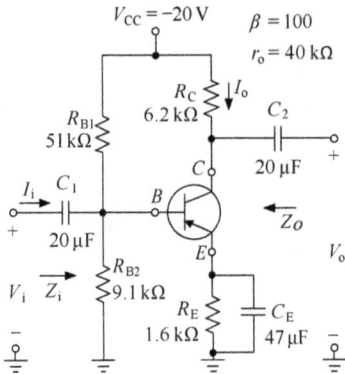

Fig. 3.58: Example 3.8.

Solution

First, obtain some AC parameters from DC analysis. As described in Section 3.3.5, check the condition to use the approximation method.

$$\beta R_E = 100 \times 1.6\,\text{k}\Omega = 160\,\text{k}\Omega$$

$$10 R_{B2} = 10 \times 9.1\,\text{k}\Omega = 91\,\text{k}\Omega$$

The following is true:

$$\beta R_E \geq 10 R_{B2}$$

Then, the approximation method can be used. So,

$$V_B = V_{CC}\frac{R_{B2}}{R_{B1} + R_{B2}} = (-20\,\text{V})\frac{9.1\,\text{k}\Omega}{9.1\,\text{k}\Omega + 51\,\text{k}\Omega} = -3.03\,\text{V}$$

Then, V_E can be calculated from the following (note that V_{BE} is negative for the *pnp* transistor):

$$V_E = V_B - V_{BE} = -3.03\,\text{V} - (-0.7\,\text{V}) = -2.33\,\text{V}$$

and the emitter current of I_E can be obtained (note that I_E flows into the *pnp* transistor):

$$I_E = \frac{-V_E}{R_E} = \frac{-(-2.33\,\text{V})}{1.6\,\text{k}\Omega} = \frac{-(-2.33\,\text{V})}{1.6\,\text{k}\Omega} = 1.46\,\text{mA}$$

Then, from Eq. (3.16),

$$r_e = \frac{26\,\text{mV}}{I_E} = \frac{26\,\text{mV}}{1.46\,\text{mA}} = 17.81\,\Omega$$

Now, AC analysis can be performed. For Z_i,

$$Z_i = R_{B1} \parallel R_{B2} \parallel \beta r_e$$

$$= (51\,\text{k}\Omega) \parallel (9.1\,\text{k}\Omega) \parallel (100 \times 1.6\,\text{k}\Omega)$$

$$= \frac{1}{\frac{1}{51\,\text{k}\Omega} + \frac{1}{9.1\,\text{k}\Omega} + \frac{1}{100 \times 1.6\,\text{k}\Omega}}$$

$$= 7.37\,\text{k}\Omega$$

For Z_o,

$$Z_o = R_C \parallel r_o = (6.2\,\text{k}\Omega) \parallel (40\,\text{k}\Omega) = 5.37\,\text{k}\Omega$$

For the voltage gain A_v,

$$A_v = \frac{-(R_C \parallel r_o)}{r_e} = -\frac{(6.2\,\text{k}\Omega) \parallel (40\,\text{k}\Omega)}{17.81\,\Omega} = -301$$

For the common-emitter configuration, no matter what type of biasing circuit used, the voltage gain A_v is always negative.

3.4.6 CE configuration with emitter bias

In Section 3.3.4, the emitter-bias circuit was discussed. Now from the viewpoint of AC analysis, the common-emitter configuration with emitter bias, as shown in Fig. 3.59, will be investigated. Substituting the r_e equivalent model into the network leads to Fig. 3.60. To simplify the analysis, the resistor r_o has not been taken into account, due to the fact that in most cases its role can be ignored.

Note that an auxiliary variable Z_b between the base and the emitter is introduced, indicating the impedance looking into the network to the right of R_B. Then, application of Kirchhoff's voltage law to the input side leads to

$$V_i = -I_b \beta r_e - I_e R_E$$

Substitute the relationship between I_e and I_b,

$$V_i = -I_b \beta r_e - I_b(\beta + 1)R_E$$

Then,

$$Z_b = \frac{V_i}{-I_b} = \beta r_e + (\beta + 1)R_E$$

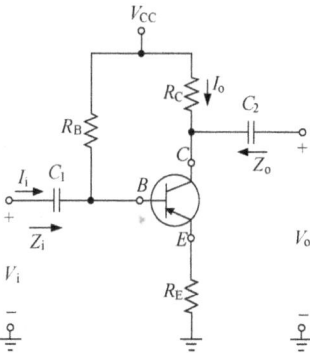

Fig. 3.59: AC parameters of the common-emitter configuration with emitter bias.

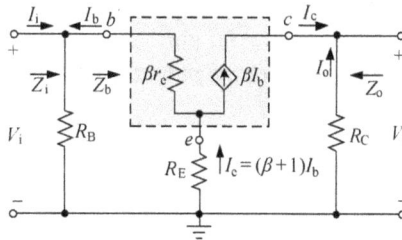

Fig. 3.60: Substituting the r_e model into the common-emitter configuration with emitter bias.

Looking into the input side, we obtain

$$Z_i = R_B \parallel Z_b$$

For Z_o, setting V_i to zero results in $I_b = 0$ and that the controlled current source can be replaced by an open-circuit equivalent. So,

$$Z_o = R_C$$

For A_v,

$$V_i = -I_b Z_b$$

and

$$V_o = (\beta I_b) R_C$$

So,

$$
\begin{aligned}
A_v &= \frac{V_o}{V_i} \\
&= \frac{(\beta I_b) R_C}{-I_b Z_b} \\
&= -\frac{\beta R_C}{Z_b} \\
&= -\frac{\beta R_C}{\beta r_e + (\beta + 1) R_E} \\
&\approx -\frac{R_C}{r_e + R_E}
\end{aligned}
$$

So, for common-emitter configuration with emitter bias,

$$A_v = -\frac{R_C}{r_e + R_E} \tag{3.22}$$

The negative sign in A_v again reveals a 180° phase shift between V_o and V_i.

Example 3.9

Fig. 3.61 shows the common-emitter configuration with emitter bias. Determine: r_e, Z_i, Z_o and A_v.

Solution

First, obtain some AC parameters from DC analysis. From Eq. (3.11) and noting that V_{BE} is negative for the *pnp* transistor

$$
\begin{aligned}
I_B &= \frac{V_{BE} - V_{CC}}{R_B + (\beta + 1) R_E} \\
&= \frac{(-0.7\,\mathrm{V}) - (-20\,\mathrm{V})}{430\,\mathrm{k\Omega} + (100 + 1) \times 510\,\Omega} \\
&= \frac{19.3\,\mathrm{V}}{430\,\mathrm{k\Omega} + 101 \times 0.51\,\mathrm{k\Omega}} \\
&= 40.08\,\mathrm{\mu A}
\end{aligned}
$$

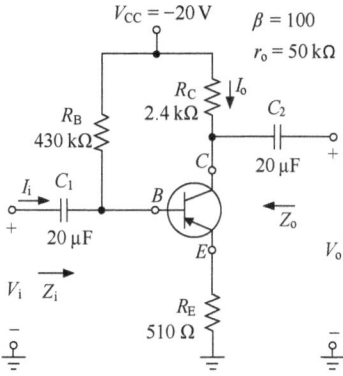

Fig. 3.61: Example 3.9.

Then,
$$I_E = (\beta + 1)I_B = (100 + 1) \times 40.08\,\mu A = 4.05\,mA$$

Then, from Eq. (3.16),
$$r_e = \frac{26\,mV}{I_E} = \frac{26\,mV}{4.05\,mA} = 6.42\,\Omega$$

Now AC analysis can be performed. For Z_i,

$$\begin{aligned}
Z_i &= R_B \,\|\, Z_b \\
&= R_B \,\|\, (\beta r_e + (\beta + 1)R_E) \\
&= 430\,k\Omega \,\|\, ((100 \times 6.42\,\Omega) + (100 + 1) \times 0.51\,k\Omega) \\
&= 430\,k\Omega \,\|\, (0.642\,k\Omega + 51.51\,k\Omega) \\
&= 430\,k\Omega \,\|\, 52.152\,k\Omega \\
&= 46.51\,k\Omega
\end{aligned}$$

For Z_o,
$$Z_o = R_C = 2.4\,k\Omega$$

For A_v,

$$\begin{aligned}
A_v &= -\frac{\beta R_C}{Z_b} \\
&= -\frac{\beta R_C}{\beta r_e + (\beta + 1)R_E} \\
&= -\frac{100 \times 2.4\,k\Omega}{(100 \times 6.42\,\Omega) + (100 + 1) \times 0.51\,k\Omega} \\
&= -\frac{240\,k\Omega}{52.152\,k\Omega} \\
&= -\frac{240\,k\Omega}{52.152\,k\Omega} \\
&= -4.60
\end{aligned}$$

As concluded before, for the common-emitter configuration, no matter what type of biasing circuit is used, the voltage gain A_v is always negative.

3.4.7 Emitter-follower configuration

In Section 3.2.3, the common-collector configuration was discussed with its DC equivalent circuit shown in Fig. 3.14. Now, its AC analysis will be investigated, as shown in Fig. 3.62.

Substituting the r_e model into Fig. 3.62 results in the network of Fig. 3.63. To simplify the analysis, the resistor r_o has been ignored.

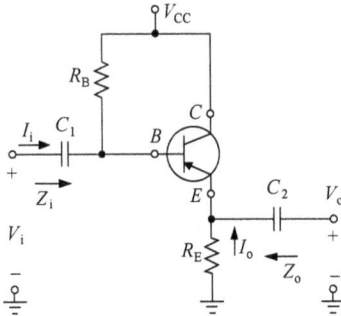

Fig. 3.62: AC parameters of common-collector configuration.

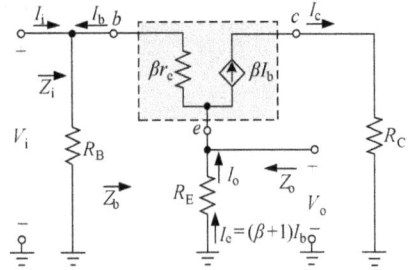

Fig. 3.63: Substituting the r_e model into common-collector configuration.

Note that an auxiliary variable Z_b between the base and emitter is introduced, indicating the impedance looking into the network to the right of R_B. Then, applying Kirchhoff's voltage law to the input side leads to

$$V_i = -I_b \beta r_e - I_e R_E$$

Substitute the relationship between I_e and I_b,

$$V_i = -I_b \beta r_e - I_b(\beta + 1)R_E$$

So,

$$Z_b = \frac{V_i}{-I_b} = \beta r_e + (\beta + 1)R_E$$

Then, looking into the input side, we obtain

$$Z_i = R_B \parallel Z_b$$

So the input impedance Z_i is obtained in the same manner as described in Section 3.4.6. Also, normally Z_i is high, as a weak load (burden) to the previous stage.

To determine the output impedance Z_o, the equation for current I_b is first obtained from the input port,

$$I_b = -\frac{V_i}{Z_b}$$

Then, substituting the relationship between I_e and I_b gives

$$I_e = (\beta + 1)I_b = -(\beta + 1)\frac{V_i}{Z_b}$$

Substituting for Z_b leads to

$$I_e = -\frac{(\beta + 1)V_i}{\beta r_e + (\beta + 1)R_E}$$

Assuming $\beta \approx \beta + 1$ results in

$$I_e \approx -\frac{V_i}{r_e + R_E}$$

Also, from Fig. 3.63,

$$V_o = -I_e R_E$$

So, substituting I_e,

$$V_o = -\left(-\frac{V_i}{r_e + R_E}\right)R_E$$

That is,

$$V_o = \frac{R_E}{r_e + R_E}V_i$$

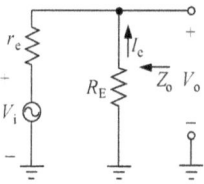

Fig. 3.64: Network to determine output impedance Z_o.

Then an equivalent circuit can be set up as shown in Fig. 3.64. By setting V_i to zero, Z_o can be determined,

$$Z_o = R_E \parallel r_e$$

Because R_E is normally much greater than r_e, the following approximation is often accepted:

$$Z_o \approx r_e$$

and r_e is normally small, so it is obvious that Z_o is low, resulting in a strong driving capability or a high output current.

Moreover, A_v can be obtained easily as the equation between V_o and V_i has been derived already:

$$A_v = \frac{V_o}{V_i} = \frac{R_E}{r_e + R_E} \tag{3.23}$$

Once again, because R_E is typically much greater than r_e, the following approximation results

$$A_v = \frac{V_o}{V_i} \approx 1 \tag{3.24}$$

The positive sign in A_v reveals that V_o and V_i are in phase; V_o is always slightly smaller than V_i. In other words, V_o always follows V_i. Additionally, based on the fact that the output V_o is taken from the emitter terminal of the transistor, the network is always referred to as the emitter-follower. However, let us not forget that it is actually the common-collector configuration. Normally, the emitter-follower configuration is used for impedance-matching purposes with its high input impedance and low output impedance.

Example 3.10
Shown in Fig. 3.65 is the common-collector configuration (emitter-follower). Determine: r_e, Z_i, Z_o and A_v.

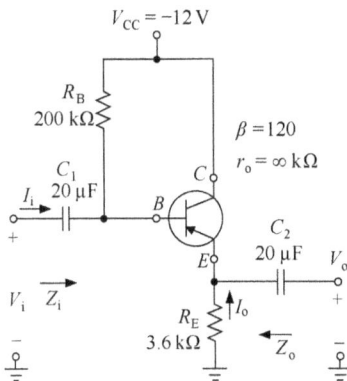

Fig. 3.65: Example 3.10.

Solution
First, obtain some AC parameters from the DC analysis. From Eq. (3.11) and noting that V_{BE} is negative for the *pnp* transistor,

$$\begin{aligned}
I_B &= \frac{V_{BE} - V_{CC}}{R_B + (\beta + 1)R_E} \\
&= \frac{(-0.7\,\text{V}) - (-12\,\text{V})}{200\,\text{k}\Omega + (120 + 1) \times 3.6\,\text{k}\Omega} \\
&= \frac{11.3\,\text{V}}{200\,\text{k}\Omega + 121 \times 3.6\,\text{k}\Omega} \\
&= 17.78\,\mu\text{A}
\end{aligned}$$

Then,

$$I_E = (\beta + 1)I_B$$
$$= (120 + 1) \times 17.78\,\mu A$$
$$= 2.15\,mA$$

Then, from Eq. (3.16),

$$r_e = \frac{26\,mV}{I_E} = \frac{26\,mV}{2.15\,mA} = 12.09\,\Omega$$

Now, AC analysis can be performed. For Z_i,

$$Z_i = R_B \parallel Z_b$$
$$= R_B \parallel (\beta r_e + (\beta + 1)R_E)$$
$$= 200\,k\Omega \parallel ((120 \times 12.09\,\Omega) + (120 + 1) \times 3.6\,k\Omega)$$
$$= 200\,k\Omega \parallel (1.45\,k\Omega + 435.6\,k\Omega)$$
$$= 200\,k\Omega \parallel 437.05\,k\Omega$$
$$= 137.25\,k\Omega$$

For Z_o,

$$Z_o = R_E \parallel r_e = 3.6\,k\Omega \parallel 12.09\,\Omega = 12.04\,\Omega$$

For A_v,

$$A_v = \frac{R_E}{r_e + R_E} = \frac{3.6\,k\Omega}{3.6\,k\Omega + 12.09\,\Omega} = 0.9967$$

As concluded before, for the emitter follower, the voltage gain A_v is always near and smaller than unity.

3.4.8 Common-base configuration

In Section 3.4.3, the r_e model for the common-base configuration was introduced. Now, by means of the r_e model, the AC analysis of the common-base configuration will be examined. The standard network of the common-base configuration, with the base terminal common to both the input and output loops of the network, is shown in Fig. 3.66. The actual directions of three currents are indicated with the applied biasing voltage, V_{EE} and V_{CC} setting up the currents. Also, six AC parameters are shown. Note that Fig. 3.66 takes the *npn* transistor as an example.

In Fig. 3.67, the r_e equivalent model is substituted into the network. The transistor output impedance r_o is ignored due to its megaohm level of value.

Then AC parameters can be found:

For Z_i, it is obvious that

$$Z_i = R_E \parallel r_e$$

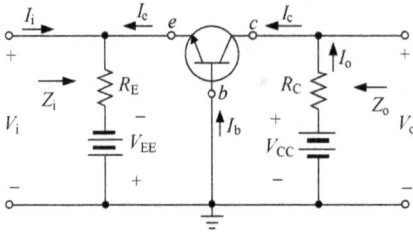

Fig. 3.66: AC parameters of the common-base configuration with *npn* transistor.

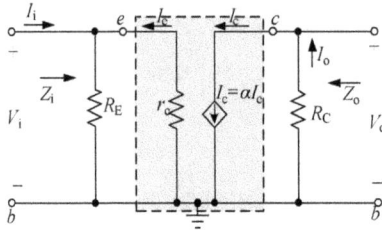

Fig. 3.67: Substituting the r_e model into the common-base configuration.

For Z_o,

$$Z_o = R_C$$

The common-base configuration is characterized as having a relatively low input and high output impedance.

In the output port, from Ohm's law, we obtain

$$V_o = -(\alpha I_e)R_C$$

and in the input port,

$$V_i = -I_e r_e$$

So, substituting I_e into V_o, we obtain

$$V_o = -\left(\alpha\left(-\frac{V_i}{r_e}\right)\right)R_C$$

Then, the voltage gain A_v is

$$A_v = \frac{V_o}{V_i} = \alpha\frac{R_C}{r_e} \approx \frac{R_C}{r_e} \tag{3.25}$$

The positive sign in A_v reveals that V_o and V_i are in phase and the voltage gain is normally quite large for the common-base configuration.

Then, for A_i, for simplicity, it is assumed that

$$R_E \gg r_e$$

So,

$$I_i = -I_e$$

and,

$$I_o = I_c = \alpha I_e$$

Then, the current gain A_i is

$$A_i = \frac{I_o}{I_i} = \frac{\alpha I_e}{-I_e} = -\alpha \approx -1 \qquad (3.26)$$

Example 3.11

Shown in Fig. 3.68 is the common-base configuration. Determine: r_e, Z_i, Z_o, A_v and A_i.

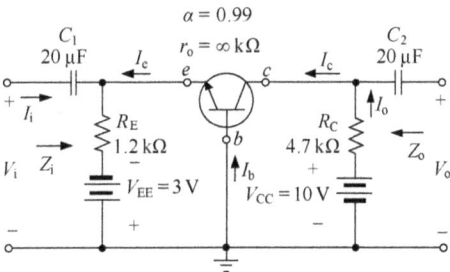

Fig. 3.68: Example 3.11.

Solution

First, obtain some AC parameters from DC analysis.

In the *b-e* loop,

$$I_E = \frac{V_{EE} - V_{BE}}{R_E} = \frac{3\,\text{V} - 0.7\,\text{V}}{1.2\,\text{k}\Omega} = 1.92\,\text{mA}$$

Then, from Eq. (3.16),

$$r_e = \frac{26\,\text{mV}}{I_E} = \frac{26\,\text{mV}}{1.92\,\text{mA}} = 13.54\,\Omega$$

Now, the AC analysis can be performed.

For Z_i,

$$Z_i = R_E \parallel r_e = 1.2\,\text{k}\Omega \parallel 13.54\,\Omega = 13.39\,\Omega$$

For Z_o,

$$Z_o = R_C = 4.7\,\text{k}\Omega$$

For A_v,

$$A_v = \frac{V_o}{V_i} \approx \frac{R_C}{r_e} = \frac{4.7\,\text{k}\Omega}{13.54\,\Omega} = 347$$

For A_i,

$$A_i = \frac{I_o}{I_i} = -\alpha \approx -0.99$$

3.5 Chapter summary

3.5.1 BJT summary

Concepts and conclusions

(1) The transistor has two types: *pnp* and *npn*. It is a three-pin device with a base terminal in the middle and an opposite type of two layers at both ends, which are the collector and the emitter. See Section 3.1.1.

The electrical symbols of *npn* and *pnp* transistors are illustrated in Fig. 3.5 as three-terminal devices. See Section 3.2.

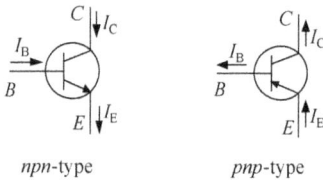

npn-type   *pnp*-type   **Fig. 3.69:** Electrical symbols of *npn* and *pnp* transistors.

(2) For DC analysis, the emitter current I_E is always the largest one, in the level of milli-ampere, whereas the base current I_B is always the smallest, in the level of micro-ampere. The emitter current I_E is always the sum of I_C and I_B. See Section 3.1.2.

(3) In the transistor symbol, the direction of the arrow defines the orientation of the emitter current I_E and, thereby, determines the direction for the other currents, I_C and I_B, and the types of the device, i.e., *npn* or *pnp*. See Section 3.2.

(4) A transistor needs two sets of characteristics to completely define its characteristics, i.e., the input and output characteristics. See Section 3.2.

(5) In the active region of the characteristics, the *b-e* junction is forward biased, whereas the *c-b* junction is reverse biased. See Section 3.3.1.

(6) For linear amplification purposes, the voltage drop across the *b-e* junction of the transistor can be assumed to be 0.7 V. See Section 3.2.1.

(7) The parameter α, relating the emitter current I_E and collector current I_C, is always close and smaller than 1. The parameter β, showing relationship between the collector current I_C and the base current I_B is usually several hundreds. See Sections 3.2.1 and 3.2.2.

Equations

$$I_E = I_C + I_B \qquad \text{Eq. (3.1)}$$

$$I_E \approx I_C \qquad \text{Eq. (3.2)}$$

$$V_{BE} = 0.7 \text{ V} \qquad \text{Eq. (3.3)}$$

$$I_C = \alpha I_E \qquad \text{Eq. (3.5)}$$
$$I_C = \beta I_B \qquad \text{Eq. (3.7)}$$
$$I_E = (\beta + 1)I_B \qquad \text{Eq. (3.10)}$$

3.5.2 DC Biasing summary

Concepts and conclusions

(1) There are three configurations for the BJT, with each pin used as common pin for both input and output loops. The analysis of *pnp* configurations is the same as for *npn* transistors, except that current directions will be opposite and voltages polarities will reverse. See Section 3.3.

(2) Under DC conditions, the operating point (*Q*-point) determines where the BJT will operate on its characteristics. For linear amplification, the DC operating point should be close to the middle of the active region, avoiding the saturation and cutoff regions. See Section 3.3.1.

(3) The desirable operating point is proper and practical DC voltages and currents, that is, (V_{CE}, I_C), based on which amplification is performed. See Section 3.3.2.

(4) For the DC analysis of a transistor network, all capacitors are replaced by an open-circuit equivalent. The DC potential is separated for both input and output loops. For most configurations, the DC analysis begins with a determination of the base current I_B. See Section 3.3.

(5) By finding the voltage and current relationships in the output or the collector loop, the equation for the load line of a transistor network can be found. The *Q*-point is then determined by finding the intersection between the load line and the base current I_B in the characteristics. This is an alternative method to determine the *Q*-point. See Section 3.3.3.

(6) For the voltage-divider bias configuration, the equivalent method is universal and can be applied to any configuration, but the approximation method is based on the assumption that I_B is zero and is much simpler while keeping an acceptable level of accuracy. See Section 3.3.5.

Equations

For fixed bias:

$$I_B = \frac{V_{BE} - V_{CC}}{R_B}$$

For the emitter-bias circuit:

$$I_B = \frac{V_{BE} - V_{CC}}{R_B + (\beta + 1)R_E}$$

For voltage-divider bias
Equivalent method:

$$I_B = \frac{V_{BE} - E_{TH}}{R_{TH} + (\beta + 1)R_E}$$

Approximation method:

$$V_B = V_{CC}\frac{R_{B2}}{R_{B1} + R_{B2}}$$

3.5.3 AC Analysis summary

Concepts and conclusions
(1) Before the AC analysis is carried out, some ideas must be clarified as follows:
   - Without the application of the DC biasing level, amplification in the AC domain is impossible.
   - The superposition theorem is used to separate the DC and AC analyses and designs for most BJT linear applications. See Section 3.4.1.
(2) Assumptions made to determine the AC model for the BJT:
   - All capacitors are replaced by a short-circuit equivalent.
   - All DC potentials are replaced by a short circuit connected to ground.
   - Rearrange the network after the introduction of short circuits or open circuits. See Section 3.4.2.
(3) There are six small-signal AC parameters, i.e., V_i, V_o, I_i, I_o, Z_i and Z_o, are used to evaluate the performance of the AC response. See Section 3.4.2.
(4) The r_e model has been introduced to the three BJT configurations. See Section 3.4.
(5) The common-emitter fixed-bias configuration can have a relatively low input impedance and a high voltage gain. The output impedance is normally assumed to be R_C. See Section 3.4.4.
(6) The common-emitter voltage-divider bias configuration is of higher stability than the fixed-bias configuration. Its voltage gain, current gain and output impedance are the same as those of the fixed-bias configuration. Due to the biasing resistors, its input impedance Z_i is lower than that of the fixed-bias configuration. See Section 3.4.5.
(7) The common-emitter emitter-bias configuration with an emitter resistor has a large input resistance Z_i and a small voltage gain A_V. The output impedance Z_o is normally R_C.
(8) The emitter-follower, actually the common-collector configuration, will always have an output signal less than and near the input signal. More importantly, the input impedance Z_i can be very large, as a weak load to extract as much of the applied signal as possible from the previous stage. Its output impedance Z_o is extremely low, leading to a strong driving capability, making it a good signal source for the following stage in a multistage system.

(9) For the common-base configuration, the input impedance Z_i is very low and the output impedance Z_o is simply R_C. However, voltage gain is significantly high while the current gain is just less than 1.

Equations

The relationship between DC and AC parameters:

$$r_e = \frac{26\,\text{mV}}{I_E}$$

Eq. (3.16)

CE configuration with fixed bias:

$$Z_i = R_B \parallel \beta r_e \approx \beta r_e$$
$$Z_o = R_C \parallel r_o \approx R_C$$
$$A_v = \frac{-(R_C \parallel r_o)}{r_e} \approx -\frac{R_C}{r_e}$$

Eq. (3.20)

CE configuration with voltage-divider bias:

$$Z_i = R_{B1} \parallel R_{B2} \parallel \beta r_e$$
$$Z_o = R_C \parallel r_o$$
$$A_v = -\frac{R_C \parallel r_o}{r_e}$$

Eq. (3.21)

CE configuration with emitter bias:

$$Z_i = R_B \parallel Z_b$$
$$Z_b = \beta r_e + (\beta + 1)R_E$$
$$Z_o = R_C$$
$$A_v = -\frac{R_C}{r_e + R_E}$$

Eq. (3.22)

Emitter-follower (common-collector configuration):

$$Z_i = R_B \parallel Z_b$$
$$Z_b = \beta r_e + (\beta + 1)R_E$$
$$Z_o \approx r_e$$
$$A_v \approx 1$$

Eq. (3.24)

Common-base configuration:

$$Z_i = R_E \parallel r_e$$
$$Z_o = R_C$$
$$A_v = \alpha \frac{R_C}{r_e} \approx \frac{R_C}{r_e}$$

Eq. (3.25)

$$A_i = -\alpha \approx -1$$

Eq. (3.26)

3.6 Questions

Q3.1: Before semiconductor transistors were invented, vacuum tubes were widely used in electronics. On the web, find the name of the vacuum tube that performs the same (or nearly the same) function as a transistor.

Q3.2: Plot the basic construction of *npn* and *pnp* transistors. Find the same and different parts between them.

Q3.3: How do majority and minority carriers in the *npn* and *pnp* transistors react to the applied biasing voltages?

Q3.4: Analyze all the biasing conditions for *npn* and *pnp* transistors. Find which biasing condition is useful in a controlled manner.

Q3.5: Draw the configurations of *npn* and *pnp* transistors. Find the difference between them.

Q3.6: Draw the input and output characteristics for different configurations of *npn* and *pnp* transistors. Find the difference between them.

Q3.7: For different configurations of *npn* and *pnp* transistors, what are the relationships between the three currents?

Q3.8: What are the conditions that should be satisfied when setting the operation point?

Q3.9: Why can the analysis of a circuit be separated as DC and AC parts?

Q3.10: What is an AC equivalent model? What conditions should it satisfy? How is it evaluated?

Q3.11: Use one (short) sentence to summarize the main purpose of DC and AC analysis.

Q3.12: What are the six parameters of AC analysis? What is the influence of each of them to the previous or following stage?

Q3.13: For the 11 examples in Chapter 3, try to change the type of device and other related parameters, and then solve the same questions again. Thus, find the same and different properties for those devices.

Q3.14: The BJT amplifier is shown in Fig. 3.70; the parameters are known: $\beta = 80$, $R_B = 510\,\mathrm{k\Omega}$, $R_C = 6.2\,\mathrm{k\Omega}$, $R_L = 7.5\,\mathrm{k\Omega}$, $V_{CC} = 10\,\mathrm{V}$, $V_{BE} = 0.6\,\mathrm{V}$. Find the

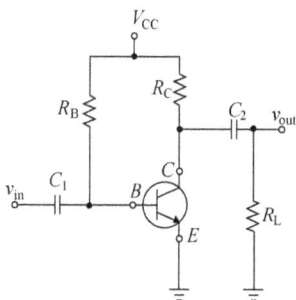

Fig. 3.70: Network of Q3.14.

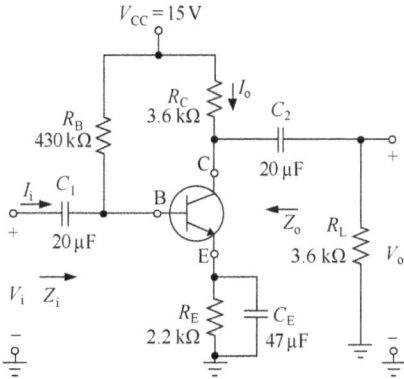

Fig. 3.71: Network of Q3.15.

following: (1) The Q-point. (2) If $I_{CQ} = 0.6$ mA and $V_{CEQ} = 5$ V are desired, find the values of R_B and R_C. (3) If the *pnp* transistor is used, which parameters should be changed for normal operation?

Q3.15: The BJT amplifier is shown in Fig. 3.71; the parameters are known: $\beta = 80$, $V_{BE} = 0.7$ V. Find the following: (1) The Q-point. (2) Find A_V, Z_i and Z_o with $r_o = 50$ kΩ. (3) If C_E is replaced with an open circuit, plot the AC equivalent circuit and find A_V, Z_i and Z_o. (4) If the *pnp* transistor is used, which parameters should be changed for normal operation?

4 Field-effect transistors

In 1925, the field-effect transistor (FET) was first patented by Julius Edgar Lilienfeld, an Austro-Hungarian-born German–American physicist and electronic engineer. Unfortunately, Lilienfeld failed to publish his articles in academic journals, and highly pure semiconductor materials were not available at that time, so his FET patent did not attract much attention and FET only existed in theory.

In 1934, Oskar Heil, the German electrical engineer and inventor, described the possibility of controlling the resistance in a semiconducting material with an electric field in a British patent.

The practical semiconductor devices JFET (junction FET) were developed in 1952, 5 years after the BJT effect was observed and explained by physicists at Bell laboratories.

In the 1950s, Martin Atalla, an Egyptian-born scientist at Bell laboratories, suggested that a FET be built of metal-oxide-silicon. He assigned the task to Dawon Kahng, a Korean-American electrical engineer in his group. At a conference in 1960, Attalla and Kahng announced their successful MOSFET (metal-oxide-semiconductor FET), which largely replaced the JFET and had great contributions to today's integrated circuits, such as microprocessors and semiconductor memory.

Generally, the field-effect transistor is a three-pin device designed for a variety of applications which, to a large extent, are the same as those accomplished by BJTs. Although there are many similarities, important differences between the two types of devices should be emphasized for an understanding of this type of device.

From the viewpoint of application, the obvious difference between the JFET and the BJT is that the JFET transistor is a voltage-controlled device, whereas the BJT is a current-controlled device, as shown in Fig. 4.1. The reader is advised to ignore the electrical symbol of JFET and terminal designators and rather focus on the controlling-controlled relationship of the variables [1, 3, 11].

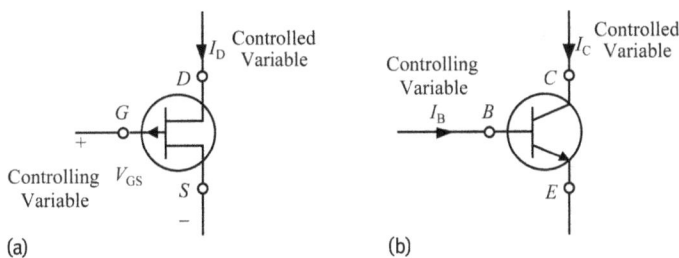

Fig. 4.1: Difference between JFETs and BJTs [11], (a) JFET, (b) BJT.

https://doi.org/10.1515/9783110593860-004

In detail, for the JFET the current I_D will be a function of the voltage V_{GS} applied to the input loop, as shown in Fig. 4.1 (a), that is,

$$I_D = f(V_{GS})$$

while the current I_C in Fig. 4.1 (b) is a direct function of the level of I_B,

$$I_C = f(I_B)$$

The exact description is given by Eq. (3.7).

Actually, the controlling and controlled quantities are connected through an electric field established by the charges present in the structure of the FET. The electric field acts in the same way as the magnetic field of the permanent magnet to attract metal filings to itself without real contact. The term "field effect" in the name of the device results from this fact.

The FET is a unipolar device that depends solely on either electrons or holes conduction. Therefore, there are n-channel and p-channel field-effect transistors, just as there are *npn* and *pnp* bipolar junction transistors.

Two main types of FETs will be introduced in this chapter: the junction field-effect transistor (JFET) and the metal-oxide-semiconductor field-effect transistor (MOSFET).

The JFET can be divided into n-channel and p-channel types. Besides n-channel and p-channel types, the MOSFET category can be further broken down into depletion and enhancement types. Therefore, there are four types of MOSFET, that is, n-channel depletion, n-channel enhancement, p-channel depletion and p-channel enhancement.

Different from the JFET, which has been normally used as independent device in the circuit, the MOSFET has been used mainly as building block of the digital integrated-circuit chips.

Moreover, there exists another type of FET, the metal-semiconductor field-effect transistor (MESFET). This is a more recent development with its advantages of high-speed performance resulting from the GaAs base material. However, it is a more expensive option in RF and computer designs. So, the balance between cost and high-speed performance should be reached. MESFET will not be discussed further in this textbook.

The advantages of FET over BJT are the following: [1, 10, 13]

(1) The FET is a unipolar device, operating on the flow of majority carriers only, while the BJT is a bipolar device, operating on the flow of both majority and minority carriers. So, the FET has less recombination noise than the BJT.

(2) The FET has high input resistance, normally in the level of MΩ, while BJT has much less input resistance. So in a multi-stage system, the FET amplifier can be regarded as a weaker load to the previous stage, if compared with the BJT amplifier. This is a very important characteristic in the design of linear AC amplifier systems.

(3) In terms of thermal stability, the FET is better than the BJT, making it particularly useful for integrated-circuit (IC) chips.
(4) The FET is relatively not easily affected by external radiation.
(5) The FET is more applicable to be implemented as a symmetrical bilateral switch.
(6) The FET is more suitable for use as a memory device on its internal capacitance or small stored charge.
(7) The FET is easier to fabricate into ICs with less space, leading to a high packing density.

The disadvantages of FET are the following: [3, 12, 16]
(1) Typically, AC voltage gains for FET amplifiers are much smaller than those for BJTs.
(2) The FET has a small gain-bandwidth product compared to the BJT. [The gain-bandwidth product for an amplifier is the product of the amplifier's bandwidth and the gain at which the bandwidth is measured. It is commonly used to determine the maximum gain that can be obtained from the amplifier for a specified frequency (or bandwidth) and vice versa.] [4]
(3) Due to the fragile insulating layer between the gate and channel, the MOSFET has a drawback of being very vulnerable to overload voltages, thus careful handling during installation is needed. This problem can be solved by proper design and installation of the circuit.
(4) FETs normally have a very low conduction resistance and a high cutoff resistance. However, the intermediate resistances are significant, which can dissipate large amounts of power while switching. Therefore, FET circuits need a trade-off between the switching speed and power dissipation. Also, a careful layout of circuits is required to avoid unintentional switching and conduction losses.

The following sections in this chapter will cover the construction, characteristics, biasing arrangements and AC analysis of both the JFET and the MOSFET.

4.1 Fundamentals of the junction field-effect transistor

4.1.1 Constructions of JFETs

As already indicated before, the JFET is a three-terminal device with one terminal in the input loop capable of controlling the current between the other two in the output loop. It's performance is the result of its specifically designed structure of semiconductor materials. The basic construction of the p-channel JFET is shown in Fig. 4.2. Note that conclusion drawn from one type of JFET will be applicable to other types with proper reversing of the polarities of voltages and directions of currents [1, 10, 12].

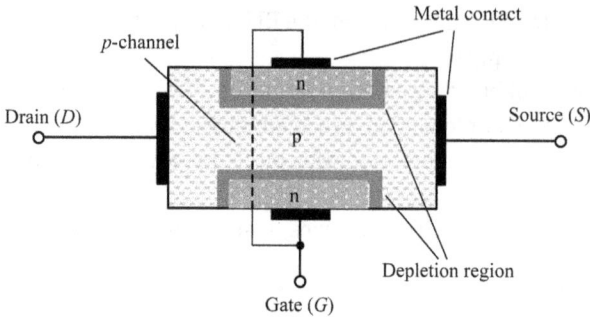

Fig. 4.2: Basic construction of the p-channel JFET.

Note that the main part of the structure is the p-type material, the lightly shadowed area in Fig. 4.2, which forms the channel between the embedded two layers of the n-type material, the heavily shadowed area in Fig. 4.2. One of the ends of the p-type channel is connected through a metal contact to an external terminal referred to as the drain (D), whereas the other end of the p-type material is connected through a metal contact to an external terminal referred to as the source (S). The two n-type materials are connected together internally and through metal contacts to the external single terminal referred to as the gate (G) [1, 10, 12].

In fact, the drain and the source are connected to the ends of the p-type channel and the gate to the two layers of the n-type material. Therefore, the JFET has two p-n junctions with one depletion region at each junction. Under the no-bias condition, that is, without any externally applied potential, the depletion region lacks free carriers and is unable to support conduction of current, the same as a diode under no-bias conditions, as shown in Fig. 4.2 [12].

The analogy between the JFET and a faucet is illustrated in Fig. 4.3. The water is pressed out of the source to flow into the drain, while the gate controls the flow. For the JFET, the applied voltage between source (S) and drain (D) establishes a flow of electrons. The gate (G), through an applied voltage, controls the flow of charge to the drain. The drain and source terminals are at either end of the p-channel, as indicated in Fig. 4.2 [10, 12].

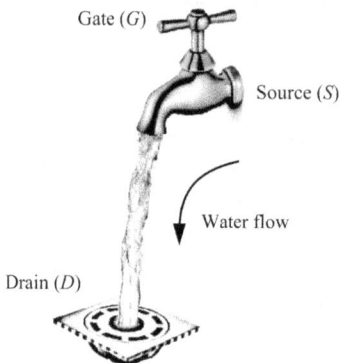

Fig. 4.3: A faucet as an analogy of the JFET.

4.1.2 Biasing conditions of JFETs

Condition: $V_{GS} = 0\,V$

In Fig. 4.4, the gate terminal is directly connected to the source terminal to establish the condition $V_{GS} = 0\,V$. A negative voltage V_{DS} is applied across the p-channel. The electrons are drawn from the source terminal, establishing the conventional current I_S with the indicated direction in Fig. 4.4. The single path of electron flow clearly verifies that the drain current I_D and source current I_S are equivalent.

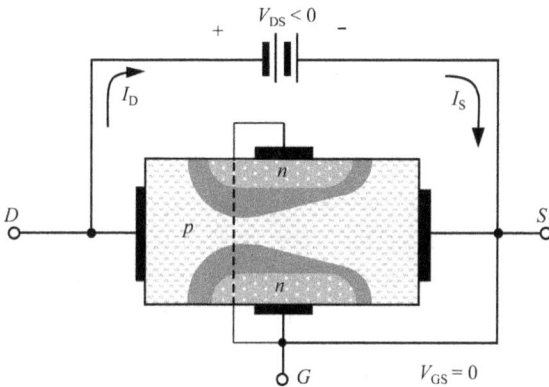

Fig. 4.4: JFET biasing when $V_{GS} = 0\,V$.

When $V_{GS} = 0\,V$, a depletion region near to the source terminal is in the no-bias condition, thus with relatively narrower width. The depletion region near the drain terminal is much wider. The flow of electrons is determined solely by the resistance of the p-channel between drain and source.

The difference in width of the depletion region can be explained as the result of the different levels of the reverse-biased condition of the p-n junction along the p-channel. Simply speaking, resistance in the p-channel can be assumed as being uniformly distributed. Along the path of the p-channel, the resistance is broken down. So, the voltage V_{DS} (or its absolute value when it is negative) drops down from full value to zero. The voltage on the gate is constant (the ground actually). So the p-n junction will be more reverse biased near the drain than near the source. Recall the conclusion of the p-n junction operation that the greater the reverse-bias voltage, the wider the depletion region. Thus, the depletion region near the drain is much wider than near the source.

More importantly, the fact that the p-n junction is reverse biased for the whole range of the p-channel results in an important characteristic of the JFET, that is, the gate current $I_G = 0\,A$.

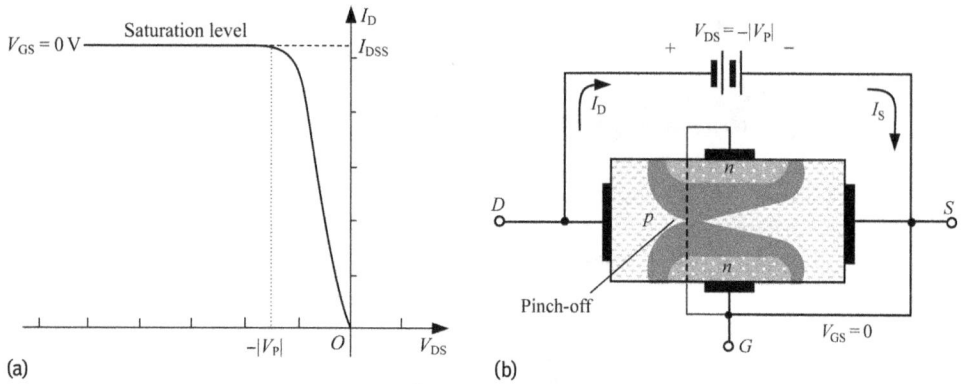

Fig. 4.5: Characteristic and state of the p-channel JFET when $V_{GS} = 0$, (a) Characteristic, (b) State when the depletion regions touch.

Moreover, as the voltage V_{DS} increases, the current I_D increases as determined by the voltage V_{DS} and the channel resistance under Ohm's law. This portion of the plot of I_D versus V_{DS} appears as the rising part of Fig. 4.5 (a). The relative straightness of the curve indicates the relative constant channel resistance. The device is in the condition that is described in Fig. 4.4.

As V_{DS} increases, the depletion regions of the device widen, leading to an obvious reduction of the channel width and an increase of the channel resistance. Furthermore, there exists a state that V_{DS} increases to a level where it appears that the two depletion regions would touch, as shown in Fig. 4.5 (b). This is the condition referred to as pinch-off. The level of V_{DS} that establishes this condition is known as the pinch-off voltage, denoted by V_P, as shown in Fig. 4.5. Actually, however, when V_{DS} reaches the pinch-off voltage V_P, I_D maintains the saturation level defined as I_{DSS} in Fig. 4.5 (a). Note that the notation I_{DSS} is based on the fact that it is the drain-to-source current with a short-circuit connection from gate to source. In fact, in the pinch-off condition, a very narrow path still exists in the p-channel, which conducts the relatively strong current of I_{DSS}.

As V_{DS} increases further beyond V_P, the touched region of the two depletion regions increases in length along the p-channel. However, the level of I_D remains essentially the same, as the horizontal part of the curve show in Fig. 4.5 (a). The current is a constant of I_{DSS}, but the voltage V_{DS} (when its absolute values $> |V_P|$) is determined by the applied load.

Note that the pinch-off voltage V_P may be originally positive or negative for different types of JFET, so in order to avoid confusion, $|V_P|$ is used to stand for positive values and $-|V_P|$ for negative values, as shown in Fig. 4.5.

Condition: $V_{GS} > 0\,V$

Now, the changes in the characteristics affected by the level of V_{GS} will be described. The voltage V_{GS} is the controlling quantity of the JFET. The curves of I_D versus V_{DS} for different levels of V_{GS} can be developed for the JFET, which is the same as the output characteristics of I_C versus V_{CE} for various levels of I_B in the analysis of the BJT.

For the p-channel device the controlling voltage V_{GS} is made more and more positive from its $V_{GS} = 0\,V$ level, that is, making the gate terminal at higher and higher potential levels as compared to the source terminal, thus keeping the depletion region along the p-channel in the reverse-biased state. In Fig. 4.6, a positive voltage of $1\,V$ is applied between the gate and source terminals.

Then, the positive V_{GS} will make I_D reach its saturation level. Moreover, as V_{GS} is made more and more positive, the resulting saturation level for I_D is reduced, as indicated in Fig. 4.7. Eventually, when $V_{GS} = |V_P|$, it will be sufficiently positive to

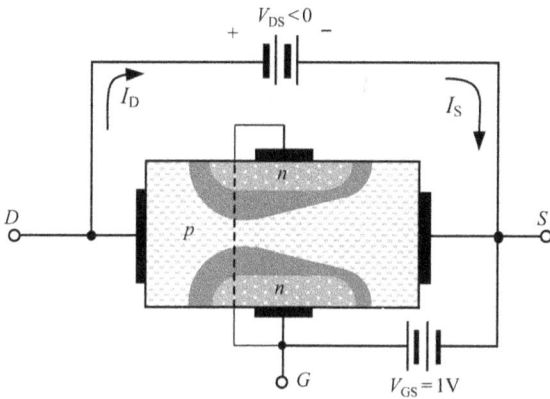

Fig. 4.6: State of the p-channel JFET when $V_{GS} > 0$.

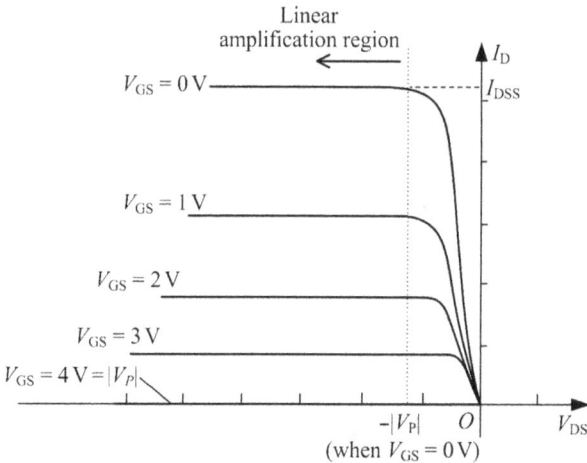

Fig. 4.7: Characteristics of p-channel JFETs.

establish a saturation level that is essentially 0 mA, and the device has actually been "turned off". On most devices' data sheets, the pinch-off voltage V_P is specified as $V_{GS(off)}$ instead.

In short, the level of V_{GS} that results in $I_D = 0$ mA is defined by $V_{GS} = V_P$, with V_P being a negative voltage for n-channel devices and a positive voltage for p-channel JFETs. In the characteristics, the region to the left of the pinch-off voltage is the one typically employed in linear amplifiers, and is commonly referred to as the linear amplification region, as illustrated in Fig. 4.7.

n-channel JFETs

The construction of n-channel JFETs is exactly the same as that of p-channel JFETs, but with opposite types of materials involved, as illustrated in Fig. 4.8.

For the n-channel device the controlling voltage V_{GS} is made more and more negative, as shown in Fig. 4.8, making I_D reach its saturation level. Also, the resulting saturation level for I_D has been reduced. Eventually, when $V_{GS} = -|V_P|$, the device has been "turned off". So, for the n-channel device the pinch-off voltage V_P is a negative voltage. In the characteristics, the region to the right of the pinch-off voltage is the linear amplification region, as illustrated in Fig. 4.9.

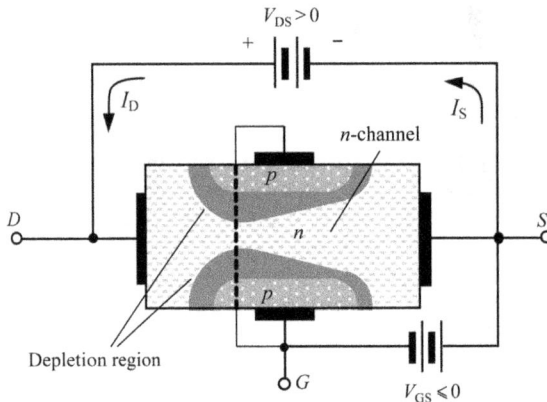

Fig. 4.8: Basic construction and biasing state of n-channel JFETs.

Symbols

The graphic symbols for n-channel and p-channel JFETs are illustrated in Fig. 4.10. The single, vertical and long bar inside the circle indicates that it is a JFET (other types of FET symbols may have more bars with different lengths). The arrow indicates the channel types. Note that the arrow always points from "P" to "N", the same as the current direction from positive potential to negative potential. So, the vertical bar with an arrow pointing inwards indicates an n-channel JFET, and the vertical bar with an arrow pointing outwards indicates p-channel JFET. Also, the current I_D is flowing into the n-channel JFET and out of the p-channel JFET.

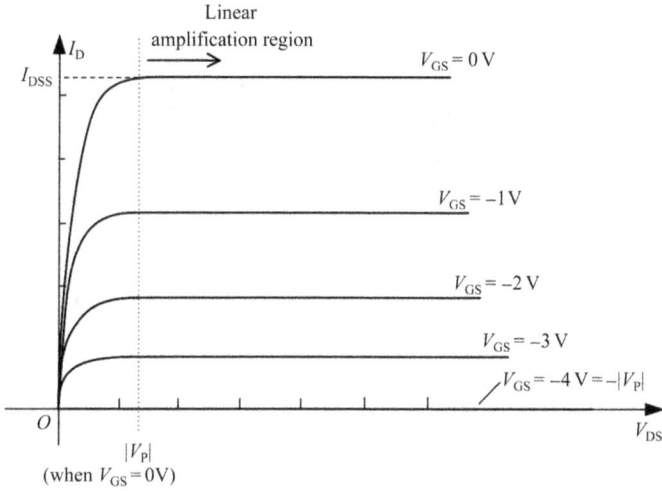

Fig. 4.9: Characteristics of *n*-channel JFETs.

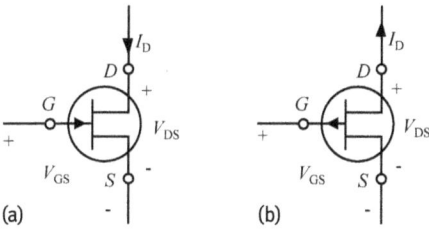

Fig. 4.10: JFET graphic symbols, (a) *n*-channel,
(b) *p*-channel.

4.1.3 JFET transfer characteristics

As indicated by Eq. (3.7), a linear relationship exists between I_C and I_B for BJTs. However, this linear relationship does not exist between the output and input quantities of a JFET. The relationship between I_D and V_{GS} is defined by Shockley's equation

$$I_D = I_{DSS} \left(1 - \frac{V_{GS}}{V_P} \right)^2 \tag{4.1}$$

The nonlinear relationship between I_D and V_{GS} is indicated clearly by the squared term in the equation, which results in an exponentially growing curve with decreasing V_{GS}. These are the transfer characteristics that are inherent and unaffected by the external network.

The transfer characteristics are of great importance to the DC analysis to be performed in the following sections. They can be obtained by direct calculation of Shockley's equation, or by other simpler methods, which will be introduced in the following discussion.

Shorthand method

Since the transfer curve will be used frequently in DC analysis, it would be quite convenient to have a shorthand method for plotting the curve in a faster, more efficient way while keeping an acceptable degree of accuracy.

The key ideal about the shorthand method is finding several points, which can be, on the one hand, easily obtained with little calculation, and on the other hand, used to plot the curve conveniently.

First, by specifying either I_D or V_{GS} zero, two points are obtained, i.e., $(0, V_P)$ and $(I_{DSS}, 0)$. The third point can be determined, if V_{GS} is set to be one-half the pinch-off value V_P, then,

$$I_D = I_{DSS} \left(1 - \frac{V_{GS}}{V_P} \right)^2$$

$$= I_{DSS} \left(1 - \frac{V_P/2}{V_P} \right)^2$$

$$= I_{DSS} \left(1 - \frac{1}{2} \right)^2$$

$$= \frac{1}{4} I_{DSS}$$

The fourth point can be obtained, if I_D is set to be one half the I_{DSS}, then,

$$I_D = I_{DSS} \left(1 - \frac{V_{GS}}{V_P} \right)^2 \triangleq \frac{I_{DSS}}{2}$$

So,

$$\left(1 - \frac{V_{GS}}{V_P} \right)^2 = \frac{1}{2}$$

Then,

$$\frac{V_{GS}}{V_P} = 1 - \sqrt{\frac{1}{2}}$$

So,

$$V_{GS} = \left(1 - \sqrt{\frac{1}{2}} \right) V_P$$

$$= (0.2928) V_P$$

$$\approx 0.3 \, V_P$$

Finally, the four points that can be used to sketch the transfer curve to a satisfactory level of accuracy are summarized in Tab. 4.1. In fact, in the DC analysis of JFETs, these points are used to sketch the transfer curves to facilitate the analysis. The transfer characteristics are given in Fig. 4.11 as the results of the shorthand method.

As illustrated in Fig. 4.11, the pinch-off voltage V_P is a negative voltage for the n-channel device and positive for the p-channel one. Note that $|V_P|$ is used to indicate positive voltage and $-|V_P|$ to indicate negative voltage, due to the fact that V_P itself would be positive or negative. The curves for the devices will be mirror images with the same limiting values.

Tab. 4.1: Four points to plot the transfer curve.

I_D	V_{GS}
I_{DSS}	0
$I_{DSS}/2$	$0.3\,V_P$
$I_{DSS}/4$	$0.5\,V_P$
0	V_P

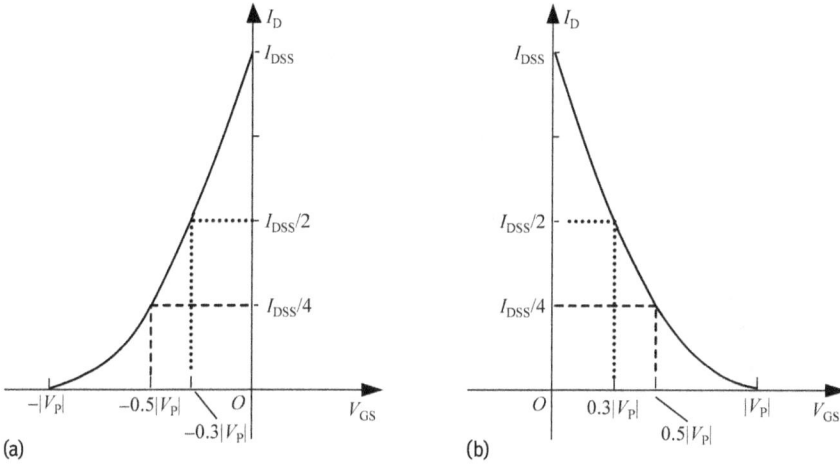

Fig. 4.11: Transfer characteristics of JFETs, (a) n-channel, (b) p-channel.

Graphical approach

The drain characteristics are shown in Figs. 4.7 and 4.9 for p-channel and n-channel JFETs, respectively. Actually, they contain the relationship of I_D versus V_{GS}, which will be illustrated clearly by the transfer characteristics. So, a graphical rather than a mathematical approach will, in general, be more direct and easier to apply, that is, obtaining transfer characteristics from drain characteristics.

Now, take the p-channel JFET as an example and redraw Fig. 4.7 as the left-hand part of Fig. 4.12. Now it is a plot of I_D versus V_{DS}, and the plot of I_D versus V_{GS} will be created. First, a new vertical axis for I_D is copied and placed to the right. The horizontal axis for V_{GS} is created in the same level as that of V_{DS}. Then, by drawing a horizontal dashed line from the flat part (saturation region) of the drain characteristics denoted by $V_{GS} = 0\,V$ to the I_D axis, a new point, $(0, I_{DSS})$, is created in this newly-created $I_D \sim V_{GS}$ coordinate system, as the point indicated by "1" in Fig. 4.12.

Then, once again in the left-hand part, from the flat part (saturation region) of the $V_{GS} = 1\,V$ drain curve, a horizontal dashed line is drawn beyond the I_D axis to the right-hand side part. On the other hand, in the right-hand side part, from the point of $V_{GS} = 1\,V$ on the abscissa, another vertical dashed line can be drawn upwards.

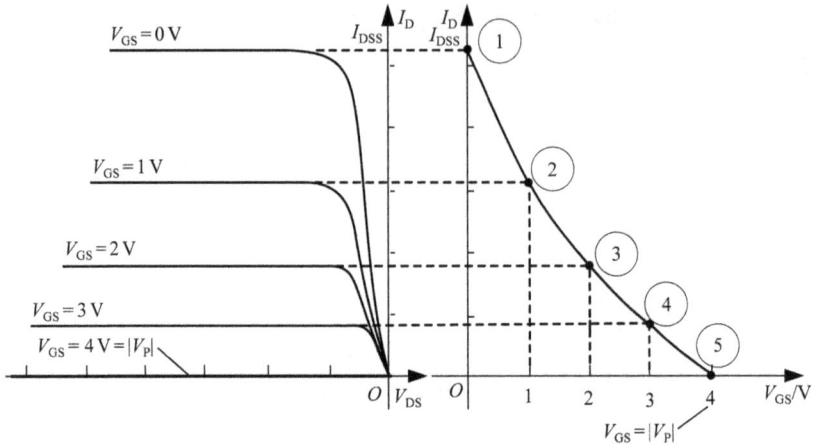

Fig. 4.12: Plotting the transfer curve from the drain characteristics.

So these two dashed line intersect, creating the second point of the transfer curve, indicated by "2" in Fig. 4.12.

Next, draw a horizontal line from the V_{GS} = 2 V drain curve beyond the I_D axis to create another intersection with a vertical line indicating V_{GS} = 2 V. This generates the third point for the transfer curve, as the one indicated by "3" in Fig. 4.12.

Repeating the same steps for the drain curve of V_{GS} = 3 V, the fourth point for the transfer curve is created, as the one indicated by "4" in Fig. 4.12.

The last point for transfer curve can be obtained in the same way, as indicated by "5" in Fig. 4.12. The only difference is that V_{GS} = V_P and the value of I_D is zero; the device is turned off.

Finally, plot a smooth curve through the five points to create the transfer characteristics, as shown in Fig. 4.12. Note that the transfer curve shows a nonlinear relationship, which can also be seen by the nonuniform spaces between the drain curves.

4.2 Fundamentals of depletion-type MOSFETs

As noted before, metal-oxide-semiconductor field-effect transistors, MOSFETs, are further broken down into depletion type and enhancement type. The terms depletion and enhancement define their basic mode of operation, which will be discussed subsequently. The difference in the characteristics and operation of those types of MOSFETs are covered in the following sections. First, the depletion-type MOSFETs will be discussed, which has characteristics similar to those of JFETs and some additional features [1, 10, 11].

4.2.1 Construction

The basic construction of the p-channel depletion-type MOSFET is illustrated in Fig. 4.13. A piece of n-type material, referred to as the substrate, is the foundation on which the device is constructed. Also, the substrate is internally connected to the source terminal, without connection to any other external terminal. The source and drain terminals are the metallic contacts, connected to p-doped regions, which are linked by a p-channel between them, as shown in Fig. 4.13. The gate is also a metal contact, connected to the surface of silicon dioxide (SiO_2) layer, but remains insulated from the p-channel. [1, 11]

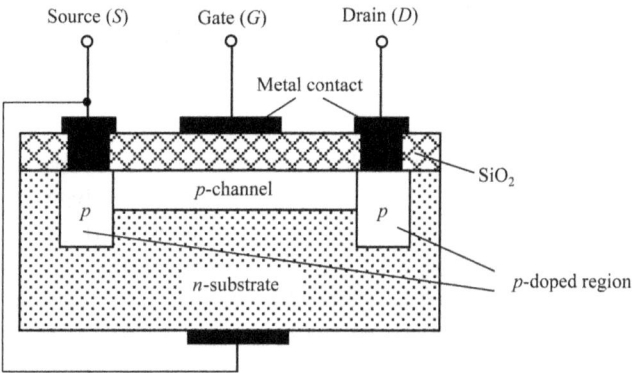

Fig. 4.13: Construction of the p-channel depletion-type MOSFET.

In fact, SiO_2, a type of insulator, is deliberately used here to set up insulation between the gate terminal and the p-channel, which leads to the fact that there is no direct electrical current through the gate terminal for DC-biased configurations. Moreover, it is the SiO_2 insulation layer in the construction that accounts for the considerably high input impedance of the MOSFET.

The n-channel depletion-type MOSFET is shown in Fig. 4.14. It is exactly the opposite of those in Fig. 4.13, that is, the substrate is now p-type and the channel is n-type. The terminals remain the same, but all the voltage polarities and the current directions will be reversed.

Now, the reason for the name of metal-oxide-semiconductor FET is rather clear: "metal" for the drain, source and gate connections to the construction; "oxide" for the silicon dioxide insulating layer, and "semiconductor" for the basic materials in which the p- and n-type regions are involved.

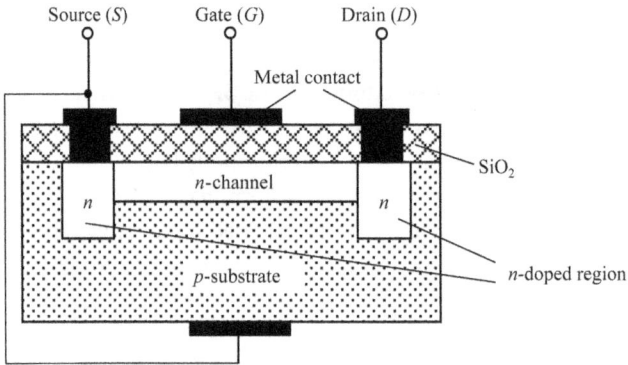

Fig. 4.14: Construction of n-channel depletion-type MOSFETs.

4.2.2 Operation and characteristics

In Fig. 4.15, the basic operation of the p-channel depletion-type MOSFET is shown. The gate and source are directly connected, leading to zero voltage difference between them. A negative voltage V_{DS} is applied across the drain-to-source terminals [1, 10, 11].

The result is an attraction of the positive particles in the p-channel by the negative terminal of V_{DS}, leading to a current, I_D or I_S through the channel. In fact, the resulting current with $V_{GS} = 0\,V$ is still labeled as I_{DSS}, as shown in Fig. 4.16.

Continue to set V_{GS} as positive voltages and the positive potential at the gate will repel the positive particles in the p-channel, resulting in a reduction of the number of positive particles in the p-channel for current conduction. When V_{GS} reaches the pinch-off voltage, I_D is reduced to zero, as shown in Fig. 4.16.

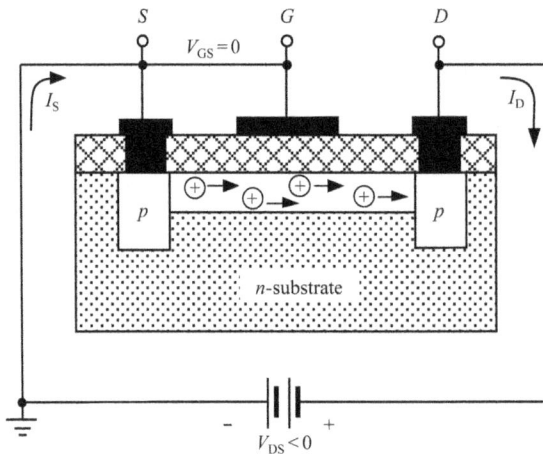

Fig. 4.15: Operation of p-channel depletion-type MOSFETs.

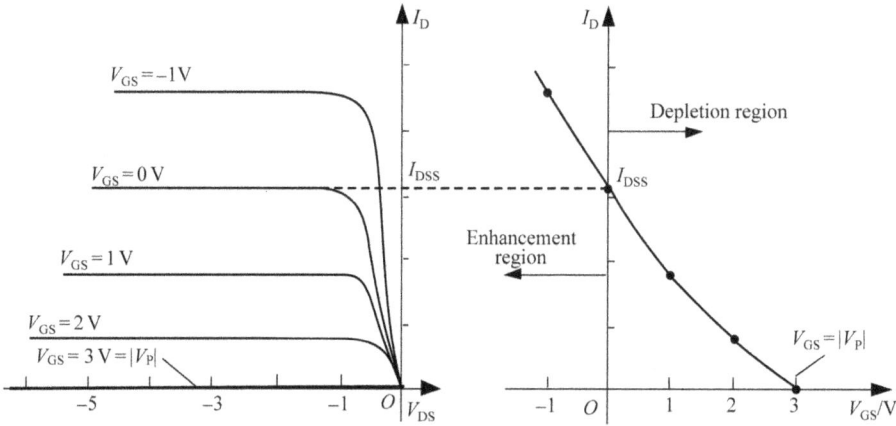

Fig. 4.16: Drain and transfer characteristics for *p*-channel depletion-type MOSFETs.

On the other hand, for negative values of V_{GS}, additional free carriers from the *n*-type substrate are attracted due to the reverse leakage current. Then, new carriers are established through the collision, resulting in accelerating particles in the current. As V_{GS} continues to be more negative, I_D will increase at a rapid rate, as shown by the transfer characteristics in Fig. 4.16. Moreover, the vertical spacing between the $V_{GS} = 0\,V$ and $V_{GS} = -1\,V$ curves of Fig. 4.16 is a clear indication of how fast the current has increased for a 1 volt change in V_{GS}. Due to the rapid rise, the user must be sure that I_D does not exceed the maximum rating when V_{GS} is negative.

As indicated above, the application of V_{GS} has "enhanced" the level of free particles in the channel compared to the condition of $V_{GS} = 0\,V$. For this reason, the region of negative V_{GS} on the transfer curve is often referred to as the "enhancement region". The region with positive V_{GS} is referred to as the "depletion region", as shown in Fig. 4.16.

More importantly, Shockley's equation, Eq. (4.1), will still be applicable for the depletion-type MOSFET characteristics in both the depletion and the enhancement regions. Note that for both regions, it is necessary that the proper signs of V_{GS} and V_P be included in the equation for mathematical operations.

n-channel depletion-type MOSFET

Briefly, for *n*-channel depletion-type MOSFETs, the drain and transfer characteristics appear in Fig. 4.17, but with V_{DS} having positive values, and V_{GS} having mostly negative values, as illustrated. From the transfer curve, it can be seen that I_D will increase from cutoff at $V_{GS} = -|V_P|$ to I_{DSS} in the negative V_{GS} region and then continue to increase for increasingly positive values of V_{GS}.

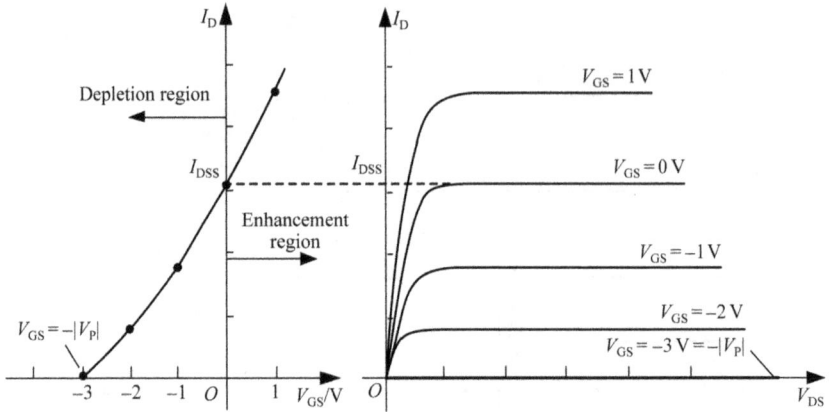

Fig. 4.17: Drain and transfer characteristics for *n*-channel depletion-type MOSFETs.

Also, the region of negative V_{GS} on the transfer curve is often referred to as the "depletion region", and that corresponding to positive V_{GS} as the "enhancement region", as shown in Fig. 4.17.

Finally, Shockley's equation, Eq. (4.1), is still applicable for the *n*-channel depletion-type MOSFET characteristics in both the depletion and the enhancement regions, with correct signs for V_{GS} and V_P in the equation.

4.2.3 Symbols

The graphic symbols for an *n*- and *p*-channel depletion-type MOSFETs are illustrated in Fig. 4.18. The actual construction of the device is reflected by the symbols. There are two separated bars between the gate and the other two terminals, indicating insulation between the gate and the channel by the SiO_2 layer. The longer vertical bar, connecting the drain and the source terminals, represents the existing semiconductors channel [1, 10, 11].

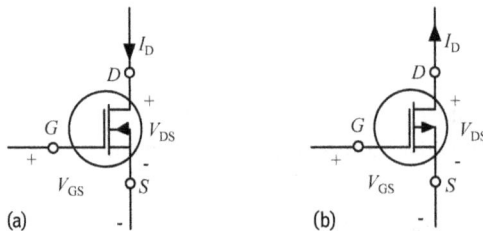

Fig. 4.18: Graphic symbols for depletion-type MOSFETs, (a) *n*-channel, (b) *p*-channel.

In the graphic symbols, the arrows indicate the channel types. Note that the arrow always points from "P" to "N", the same as the current direction from positive potential to negative potential. So, the longer vertical bar with an arrow pointing inwards indicates an n-channel JFET, and the longer vertical bar with an arrow pointing outwards indicates p-channel JFET. Also, the current I_D flows into the n-channel with positive V_{DS} and out of the p-channel with negative V_{DS}.

Also, two symbols reflect the fact that the substrate is internally connected to the source terminal as a three-terminal device. This is the type of device that will be discussed throughout the text.

4.3 Fundamentals of enhancement-type MOSFETs

Now, it is time for the discussion of enhancement-type MOSFETs. The characteristics of enhancement-type MOSFETs are quite different from the conclusions obtained thus far for depletion-type MOSFETs, although some similarities in construction and operation exist between them. Moreover, another obvious difference would be that the transfer characteristics of enhancement-type MOSFETs are not defined by Shockley's equation, and I_D only exists when V_{GS} reaches a specific level.

4.3.1 Construction

The basic construction of the p-channel enhancement-type MOSFET is illustrated in Fig. 4.19. A piece of n-type material, referred to as the substrate, is the foundation on which the device is constructed. Also, the substrate is internally connected to the source terminal, without connection to any other external terminal. The source and drain terminals are the metallic contacts, connected to p-doped regions. Moreover, between them, there is no channel as a constructed component of the device, as illustrated in Fig. 4.19. This is the most obvious difference in construction between depletion-type and enhancement-type MOSFETs. The gate is also a metal platform, connecting to the surface of silicon dioxide (SiO_2) layer between the source and drain, and still insulated from the n-substrate [11–13, 16].

The same as for depletion-type MOSFETs, the deliberately-designed SiO_2 layer can set up insulation between the gate terminal and the n-substrate, resulting in no electrical current through the gate terminal. Moreover, this also accounts for the high input impedance of the MOSFETs.

Therefore, in short, the construction of an enhancement-type MOSFET is almost the same as that of the depletion-type MOSFET, except for the absence of a channel between the source and drain terminals.

For n-channel enhancement-type MOSFETs, the construction is illustrated in Fig. 4.20. It is exactly the opposite of that in Fig. 4.19, that is, the type of semiconduc-

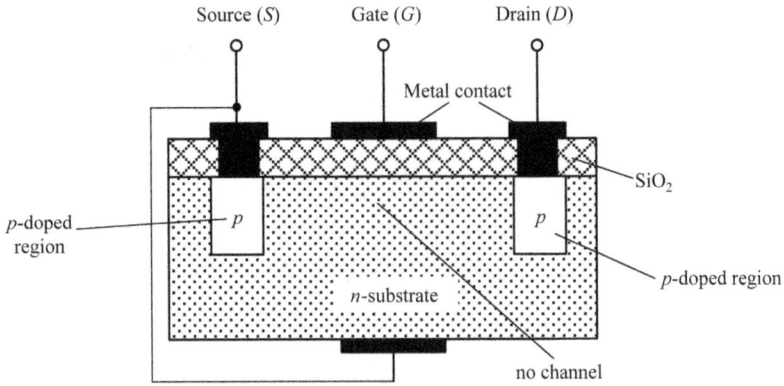

Fig. 4.19: Construction of p-channel enhancement-type MOSFETs.

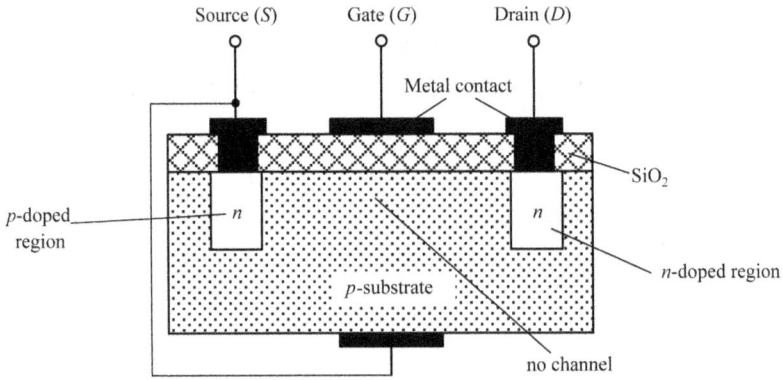

Fig. 4.20: Construction of n-channel enhancement-type MOSFETs.

tor materials involved changes. The terminals remain the same, but all the voltage polarities and the current directions are reversed [11, 12].

4.3.2 Operation and characteristics

With the basic construction of p-channel enhancement-type MOSFETs shown in Fig. 4.19, if V_{GS} is set at zero and V_{DS} is some values of the voltage, there will be no current through the gate or drain, due to the fact that no channel exists between them.

In Fig. 4.21, both V_{DS} and V_{GS} have been set at some negative voltages, making the drain and the gate at lower potential with respect to the source. The negative potential at the gate terminal will pressure the electrons (negative particles) in the n-substrate away from the border of the SiO_2 layer to enter deeper into the n-substrate region, as shown in Fig. 4.21 [11, 16].

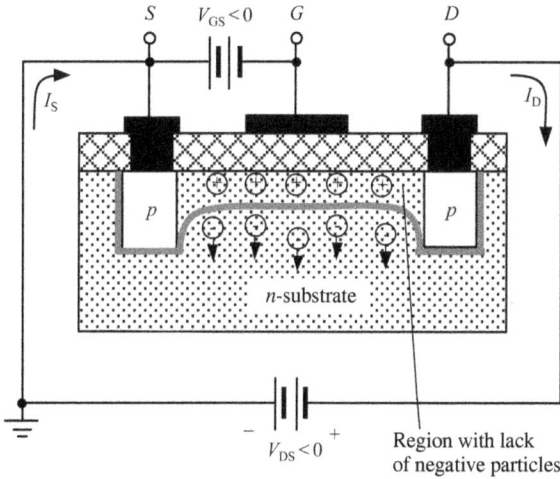

Fig. 4.21: Process of channel enhancement in the *p*-channel enhancement-type MOSFET.

The consequence is the generation of a depletion region along the SiO_2 insulation layer with a lack of electrons. On the other hand, the minority carriers of the *n*-substrate, the holes, will be attracted to the negative gate and accumulate in the region along the border of the SiO_2 layer. The SiO_2 layer will definitely prevent the positive particles from being absorbed through the gate terminal. However, as V_{GS} increases in magnitude, the accumulation of holes near the SiO_2 border also increases until, finally, the induced *p*-type channel can support a measurable current of I_D. The level of V_{GS} that leads to the significant increase in I_D is called the threshold voltage, denoted by V_T, which will be used frequently in the following analysis. On some data sheets, V_T is also referred to as $V_{GS(Th)}$.

Since an induced *p*-type channel does not exist when $V_{GS} = 0\,V$ and "enhanced", then by the application of a negative V_{GS}, this type of MOSFET is named an enhancement-type MOSFET.

As V_{GS} becomes more negative than V_T, the density of holes in the enhanced p-channel will increase, leading to an increased level of I_D. However, if V_{DS} increases in the condition of constant V_{GS}, I_D will eventually level off at its maximum value, due to the pinch-off condition resulting from the narrow width of the enhanced channel, as shown in Fig. 4.22. The different width of the enhanced *p*-channel is the consequence of the different voltage drop along the channel, which generates a depletion region between the *p*-doped regions and the *n*-substrate.

For levels of $|V_{GS}| < V_T$, the drain current of I_D is zero, and for levels of $|V_{GS}| > V_T$, I_D is related to the applied V_{GS} by the following nonlinear relationship:

$$I_D = k\,(V_{GS} - V_T)^2 \tag{4.2}$$

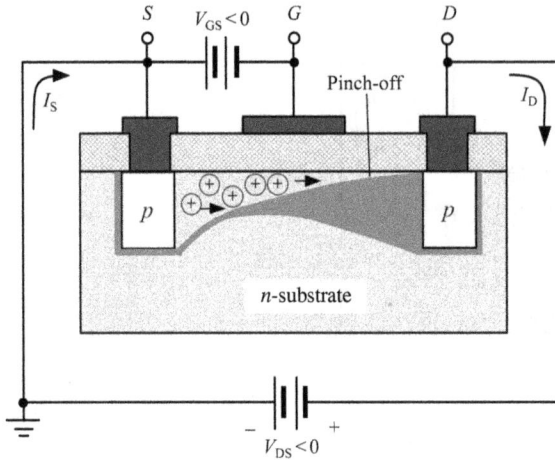

Fig. 4.22: Pinch-off condition of the enhanced channel.

where k is a constant determined by the construction of the device. It can be obtained from the following equation:

$$k = \frac{I_{D(on)}}{\left(V_{GS(on)} - V_T\right)^2} \qquad (4.3)$$

where $I_{D(on)}$ and $V_{GS(on)}$ are the parameters at a particular point on the device characteristics. Normally, they can be obtained from the data sheets.

Just like Shockley's equation of Eq. (4.1), Eq. (4.2) is the squared term that results in the nonlinear relationship between I_D and V_{GS}.

The drain characteristics of the p-channel enhancement-type MOSFET, I_D versus V_{GS}, are shown in the left-hand side of Fig. 4.23. Note that both I_D and V_{GS} are negative for p-channel devices, and I_D begins only after V_{GS} is greater than V_T.

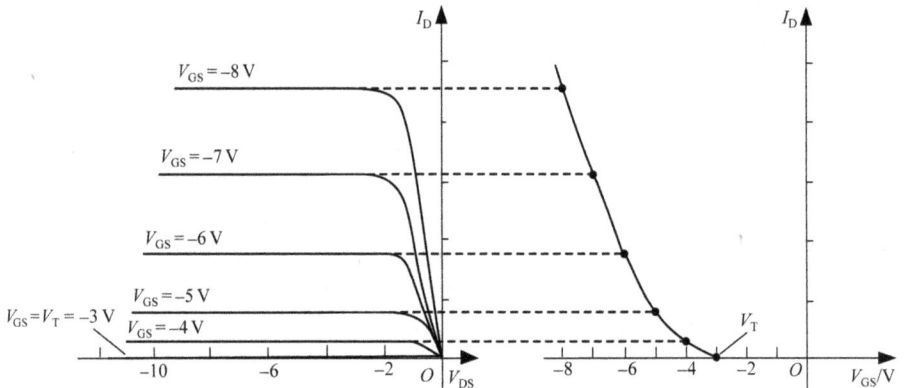

Fig. 4.23: Drain and transfer characteristics for p-channel enhancement-type MOSFETs.

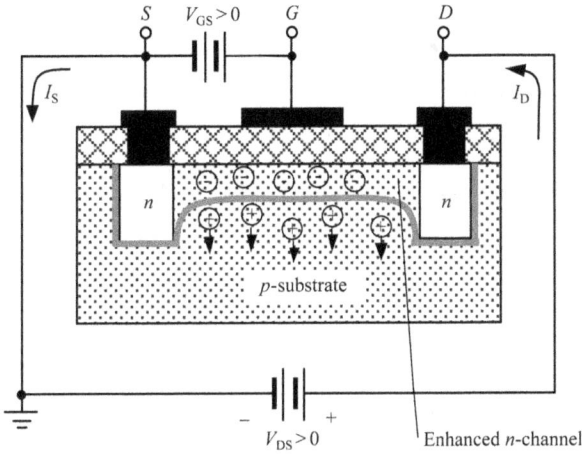

Fig. 4.24: Operation of the *n*-channel enhancement-type MOSFET.

The transfer characteristics of enhancement-type MOSFETs are also important for the DC analysis and they will be obtained through the graphical method as illustrated in Fig. 4.23. The drain characteristics are first drawn on the left-hand side. Then the transfer characteristics are set side by side to demonstrate the transfer process point by point. More importantly, the drain current I_D only exists when V_{GS} is greater than V_T. In this case, I_D will be defined by Eq. (4.2). Note that in the process of determining the points on the transfer characteristics from the drain characteristics, only the saturation currents are employed, thereby confining V_{GS} within the region that is greater than V_T. Actually, this method is the same as those for the JFET and depletion-type MOSFETs in the previous sections in drawing transfer characteristics.

Now, the operations of *n*-channel enhancement-type MOSFET will be discussed. The operation of an *n*-channel enhancement-type MOSFET is exactly the reverse of that appearing in Fig. 4.21, as shown in Fig. 4.24. It can be seen that there are a *p*-type substrate and two *n*-doped regions under the drain and source terminals. The terminal notations are same as those of depletion-type MOSFETs. However, now V_{GS} and V_{DS} are positive, and I_D and I_S are the reverse of those of depletion-type MOSFETs [13, 16].

The drain characteristics of *n*-channel enhancement-type MOSFETs are shown in the right-hand side of Fig. 4.25, with positive values of V_{DS} in different conditions of positive values of V_{GS}. Also, the drain current I_D only exists when V_{GS} is greater than V_T. The transfer characteristics, shown in the left-hand side of Fig. 4.25, are plotted by the graphical method from drain characteristics.

Now, it is clearer that I_D will have measurable values when V_{GS} is greater than V_T. The calculation of I_D can be performed by Eq. (4.2) for *n*-channel devices, the same as for *p*-channel devices.

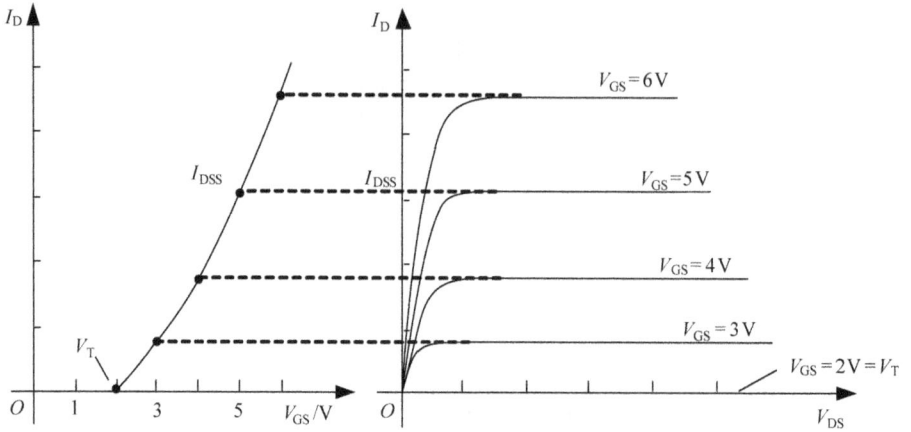

Fig. 4.25: Drain and transfer characteristics for n-channel enhancement-type MOSFETs.

4.3.3 Symbols

The graphic symbols for an n- and p-channel enhancement-type MOSFETs are illustrated in Fig. 4.26 [11–13, 16]. The actual construction of the device is reflected by the symbols. For both symbols, the substrate is internally connected to the source terminal as a three-terminal device. There are two separated bars between the gate and the other two terminals, indicating insulation between the gate and the channel by the SiO_2 layer. The longer vertical bar, now shown with a dashed line for the enhancement-type device, connecting the drain and the source terminals, represents the enhanced channel. Also, in the graphic symbols, the arrows indicate the channel types. Note that the arrow always points from "P" to "N", the same as the current direction from positive potential to negative potential. So, the bar pointed to by an arrow indicates the n-channel and the bar with an arrow pointing away indicates the p-channel. Also, the current I_D flows into the n-channel with positive V_{DS} and out of the p-channel with negative V_{DS}.

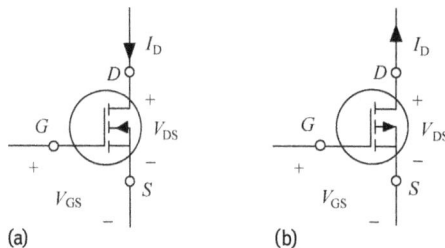

Fig. 4.26: Graphic symbols for enhancement-type MOSFETs, (a) n-channel, (b) p-channel.

4.4 JFET DC biasing configurations

In Chapter 3, Eq. (3.7), repeated as follows:

$$I_C = \beta I_B$$

is the most important relationship to obtain the biasing voltages and currents of BJT configurations. The link between input and output parameters is described by β, which is normally assumed as a constant in the DC analysis, establishing a linear relationship [3, 11].

However for JFETs, due to the squared term in Shockley's equation of Eq. (4.1), the relationship between I_D and V_{GS} is nonlinear, thus complicating the mathematical approach to the DC analysis of JFET configurations. So, the graphical approaches, which are quicker and more convenient than direct mathematical ones, will be used for the analysis for the JFET amplifier, with sufficiently good levels of accuracy.

Another main difference of the analysis between BJTs and JFETs is that the input controlling parameter for the BJT is the current of I_B, whereas the controlling parameter for the JFET is the voltage of V_{GS}. The controlled variable on the output side for the BJT is the current of I_C, whereas the controlled quantity on the output side for the JFET is the current of I_D.

The relationships that can be applied to the DC analysis of all JFET amplifiers are

$$I_D = I_S \tag{4.4}$$

and

$$I_D \approx 0 \, \text{A} \tag{4.5}$$

Once again, Shockley's equation relates the input and output variables of JFETs [3, 11, 12, 16]:

$$I_D = I_{DSS} \left(1 - \frac{V_{GS}}{V_P} \right)^2$$

The same as for the BJT, the DC solution for JFET networks is the solution of simultaneous equations established by the device and the network, the former is transfer characteristics and the latter is the voltage and current relationships from the circuits.

4.4.1 Fixed-bias configuration

The fixed-bias configuration for the p-channel JFET appears in Fig. 4.27, which is the simplest one among different biasing arrangements of JFET. However, because the configuration needs two DC supplies, its application is limited and does not belong to the category of popular JFET configurations [3, 11].

The JFET fixed-bias configuration includes the AC signal terminals for v_{in} and v_{out}, connected with coupling capacitors, C_1 and C_2. The same as discussed previously

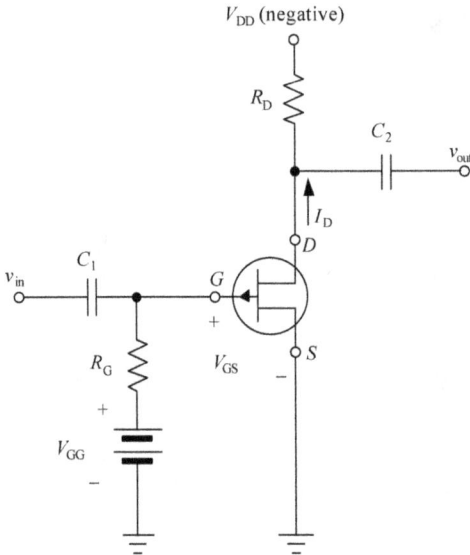

Fig. 4.27: Fixed-bias configuration for *p*-channel JFETs.

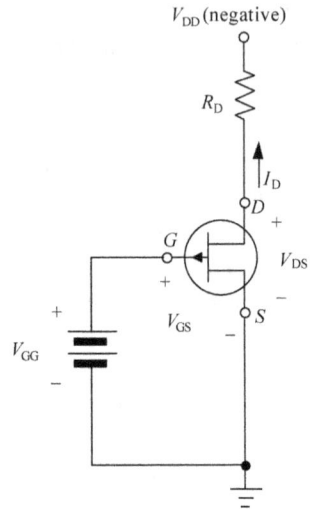

Fig. 4.28: DC equivalent circuit of fixed-bias configuration for *p*-channel JFETs.

for BJT configurations, the coupling capacitors are "open circuits" due to their high impedances for the DC analysis and "short circuits" due to their low impedances for the AC analysis, as shown in Fig. 4.28. The resistor R_G is present to ensure that v_{in} can be fed to the input terminal of the JFET amplifier for the AC analysis, which will be a discussion focus in the subsequent sections.

For the DC analysis of all JFETs, as in Eq. (4.5),

$$I_D \approx 0 \, A$$

which will be the starting point for the DC analysis. Then the voltage drop across R_G will be zero volt, leading to R_G being replaced by a short-circuit equivalent, as appears in the DC equivalent circuit of Fig. 4.28.

The JFET fixed-bias configuration is so straightforward that either the mathematical or the graphical approach can be used to find the solution of the DC analysis.

Graphical approach

The transfer characteristics of JFET, as drawn from Shockley's equation, Eq. (4.1), are necessary in the graphical approach. For simplicity, the shorthand method, as introduced in Section 4.1.3, can be used to plot the curve by the four points summarized in Tab. 4.1. They are illustrated in Fig. 4.29.

From the DC equivalent circuit of Fig. 4.28, we obtain that

$$V_{GS} = V_{GG}$$

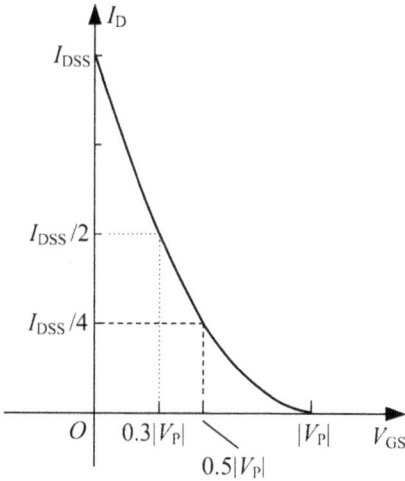

Fig. 4.29: Transfer characteristics of p-channel JFETs drawn by the shorthand method.

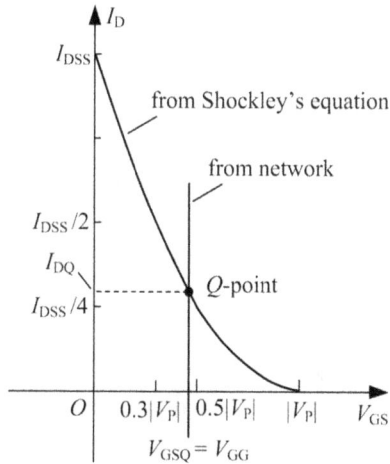

Fig. 4.30: Q-point obtained by the graphical approach.

It is actually a vertical straight line when V_{GS} is regarded as a function, as shown in Fig. 4.30. Moreover, since V_{GG} is a constant DC power supply, the fixed voltage magnitude of V_{GG} accounts for the designation of "fixed-bias configuration".

So, the two curves are ready, one from the device itself, described as Shockley's equation, the other from the network, the DC equivalent circuit. Then, the point where the two curves intersect is the common solution to the configuration, the same as for BJTs, referred to as the quiescent point, or simply the Q-point.

Then from the Q-point just obtained, by drawing a horizontal line to the vertical I_D axis, the quiescent value of I_D is determined, which is indicated by I_{DQ}. The quiescent value of V_{GS} for the Q-point is obviously V_{GG}, specifically denoted by V_{GSQ}. The subscript Q will specifically be applied to identify their values at the Q-point.

The voltage of V_{DS} in the output loop can be determined as follows:

$$V_{DS} = V_{DD} + I_D R_D$$

Note that V_{DD} and V_{DS} are both negative for the p-channel JFET. The remaining work of finding the values of V_G, V_S and V_D will not be difficult.

Mathematical approach
From the DC equivalent circuit of Fig. 4.28, we obtain

$$V_{GS} = V_{GG}$$

The values of the output quantity, the drain current I_D, is now determined by Shockley's equation:

$$I_D = I_{DSS}\left(1 - \frac{V_{GS}}{V_P}\right)^2$$

So, simply substitute V_{GS}, including its magnitude and sign, into Shockley's equation and the resulting value of I_D can be calculated. The remaining work of finding the values of V_{DS}, V_G, V_S and V_D will be exactly the same as for the graphical approach. For JFET fixed-bias configuration, the mathematical approach is rather direct and straightforward.

Example 4.1

The p-channel JFET fixed-bias configuration is shown in Fig. 4.31. Find the values of V_{GSQ}, I_{DQ}, V_{DS}, V_D, V_G and V_S for the network.

Solution

First, change the JFET fixed-bias configuration to its DC equivalent circuit, as shown in Fig. 4.32. Note that R_G, C_1 and C_2 are removed.

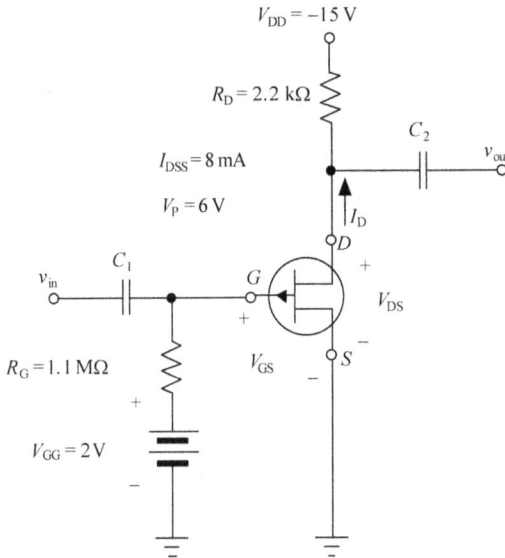

Fig. 4.31: Network of Example 4.1.

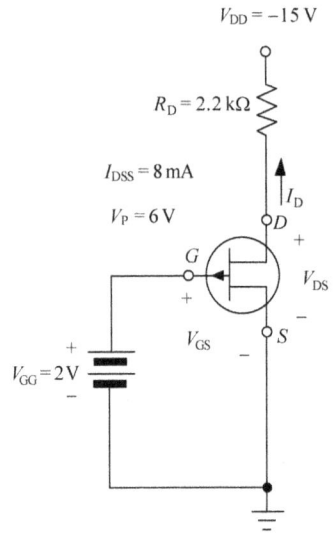

Fig. 4.32: DC equivalent circuit of Example 4.1.

Graphical approach

The four points for the shorthand method are summarized in Tab. 4.2.

Tab. 4.2: Four points to plot the transfer curve of Example 4.1.

Points	Values
$(0, I_{DSS})$	$(0\,V, 8\,mA)$
$(0.3\,V_P, I_{DSS}/2)$	$(1.8\,V, 4\,mA)$
$(0.5\,V_P, I_{DSS}/4)$	$(3\,V, 2\,mA)$
$(V_P, 0)$	$(6\,V, 0\,mA)$

Then the transfer curve can be plotted, as illustrated in Fig. 4.33.
From the DC equivalent circuit of Fig. 4.32, we obtain that

$$V_{GS} = V_{GG} = 2\,V$$

This is actually a vertical straight line. So, the two curves are ready, one from the device itself, the other from the network. Then, the solution, the Q-point, is just the intersection of the two curves, as shown in Fig. 4.34. So, from Fig. 4.34, it is very clear that

$$V_{GSQ} = 2\,V$$

$$I_{DQ} \approx 3.6\,mA$$

Then from the output loop of Fig. 4.32, V_{DS} can be determined as follows:

$$V_{DS} = V_{DD} + I_D R_D = -15\,V + (3.6\,mA \times 2.2\,k\Omega) = -7.08\,V$$

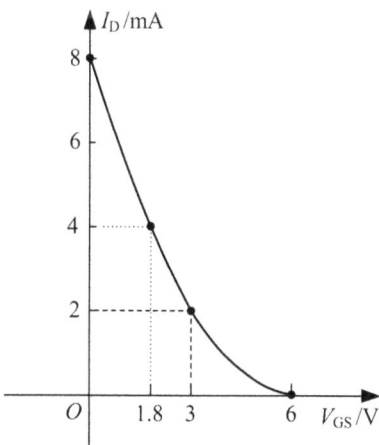

Fig. 4.33: Transfer characteristics drawn by the shorthand method for Example 4.1.

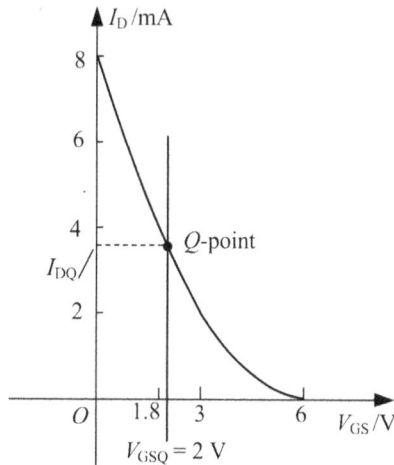

Fig. 4.34: Q-point obtained by the graphical approach for Example 4.1.

For V_D

$$V_D = V_{DS} = -7.08\,\text{V}$$

For V_G

$$V_G = V_{GS} = V_{GG} = 2\,\text{V}$$

For V_S

$$V_S = 0\,\text{V}$$

Mathematical approach

In the output loop of Fig. 4.32, the drain current I_D is determined by Shockley's equation. By substituting $V_{GSQ} = 2\,\text{V}$ and $V_P = 6\,\text{V}$ into it, the resulting value of I_D can be calculated:

$$I_D = I_{DSS}\left(1 - \frac{V_{GS}}{V_P}\right)^2$$

$$= 8\,\text{mA}\left(1 - \frac{2\,\text{V}}{6\,\text{V}}\right)^2$$

$$= 8\,\text{mA}\left(\frac{2}{3}\right)^2$$

$$= 3.56\,\text{mA}$$

It can be seen that the two magnitudes of I_D from either the graphical or the mathematical method are so close that they can be regarded as having the same level of accuracy. The remaining work of finding the values of V_{DS}, V_G, V_S and V_D will be exactly the same as for the graphical approach.

4.4.2 Self-bias configuration

Unlike fixed-bias configuration, which needs two DC supplies, self-bias configuration uses only one single DC power. The controlling parameter V_{GS} is now obtained through the resistor R_S connected to the terminal source, as shown in Fig. 4.35.

The JFET self-bias configuration also includes the AC signal terminals for v_{in} and v_{out}, connected with coupling capacitors, C_1 and C_2. For the DC analysis, these capacitors can be replaced by "open circuits". The resistor R_G in the input loop will be replaced by a short-circuit equivalent due to that $I_D = 0\,\text{A}$, although its role in the AC analysis is important to feed v_{in} into the JFET amplifier. The DC equivalent circuit of self-bias configuration is shown in Fig. 4.36.

The current through R_S is the source current I_S, and it is obvious that $I_S = I_D$. In the input loop, we obtain

$$V_{GS} = I_S R_S$$

Using the relationship of $I_S = I_D$,

$$V_{GS} = I_D R_S$$

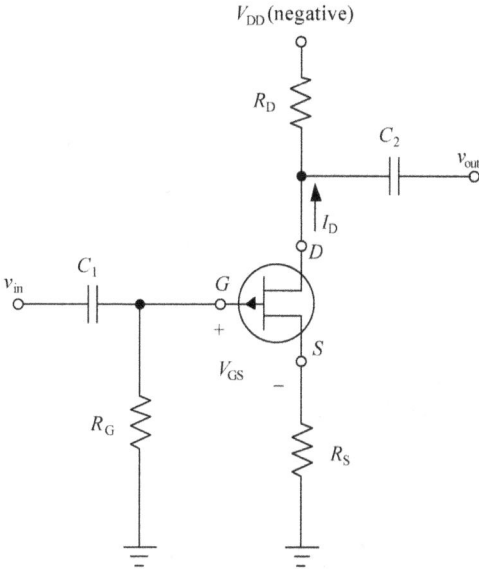

Fig. 4.35: Self-bias configuration for the p-channel JFET.

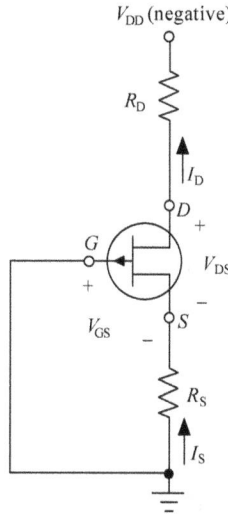

Fig. 4.36: DC equivalent circuit of self-bias configuration for the p-channel JFET.

So,

$$I_D = \frac{1}{R_S} V_{GS}$$

Note that I_D is a function of V_{GS} and it is a straight line through the origin.

Then by the graphical approach, transfer characteristics are first established by four points. Since the $I_D \sim V_{GS}$ function defines a straight line through the origin, one more point is necessary to draw it. Any applicable values of V_{GS} can be used to determine the second point, for example, the point of $((R_S I_{DSS})/2, I_{DSS}/2)$, as illustrated in Fig. 4.37.

So, the two relationships are ready, the transfer curve and the straight line. Then, the point where the two curves intersect is the solution to the DC analysis, the Q-point. By drawing a horizontal line from the Q-point to the vertical I_D axis, I_{DQ} is determined. Further, by drawing a vertical line from the Q-point to the V_{GS} axis, V_{GSQ} is obtained, as illustrated in Fig. 4.37.

The voltage of V_{DS} in the output loop, as shown in Fig. 4.36, can be determined as follows:

$$V_{DS} = V_{DD} + I_D(R_D + R_S)$$

Note that V_{DD} and V_{DS} are both negative for the p-channel JFET. The remaining work of finding the values of V_G, V_S and V_D will be easy.

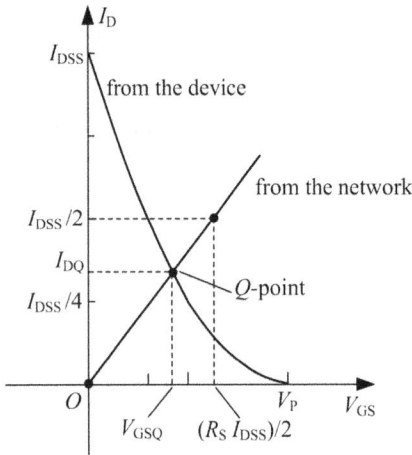

Fig. 4.37: Q-point obtained by the graphical approach for self-bias configuration of the p-channel JFET.

Example 4.2

The p-channel JFET self-bias configuration is shown in Fig. 4.38. Find the values of V_{GSQ}, I_{DQ}, V_{DS}, V_D, V_G and V_S for the network.

Solution

First, convert the JFET self-bias configuration to its DC equivalent circuit, as shown in Fig. 4.39. Note that R_G, C_1 and C_2 are removed.

The four points for the shorthand method to plot transfer curve are summarized in Tab. 4.3.

Then the transfer curve can be plotted, as illustrated in Fig. 4.40.

For the curve of $I_D \sim V_{GS}$, the point of $((R_S\,I_{DSS})/2, I_{DSS}/2)$ will be used. Substituting the values of I_{DSS} and R_S, we obtain

$$\frac{R_S \cdot I_{DSS}}{2} = \frac{1.1\,\text{k}\Omega \times 10\,\text{mA}}{2} = 5.05\,\text{V}$$

and

$$I_{DSS}/2 = 5\,\text{mA}$$

So the point (5.05 V, 5 mA) is obtained and the straight line can be determined. Then, the solution to the DC analysis, the Q-point, can be obtained by finding the intersection of the two curves, the transfer curve and the straight line, as shown in Fig. 4.40.

So, from the Q-point in Fig. 4.40, it's very obvious that

$$V_{GSQ} \approx 3.1\,\text{V}$$
$$I_{DQ} \approx 3.0\,\text{mA}$$

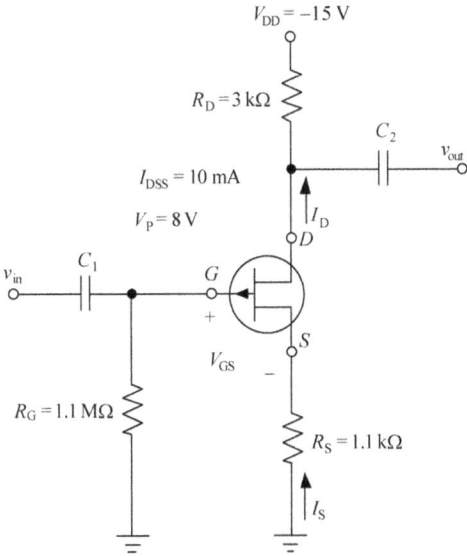

Fig. 4.38: Network of Example 4.2.

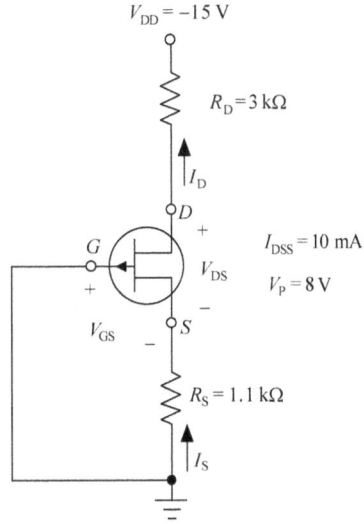

Fig. 4.39: DC equivalent circuit of Example 4.2.

Tab. 4.3: Four points to plot transfer curve of Example 4.2.

Points	Values
$(0, I_{DSS})$	$(0\,V, 10\,mA)$
$(0.3\,V_P, I_{DSS}/2)$	$(2.4\,V, 5\,mA)$
$(0.5\,V_P, I_{DSS}/4)$	$(4\,V, 2.5\,mA)$
$(V_P, 0)$	$(8\,V, 0\,mA)$

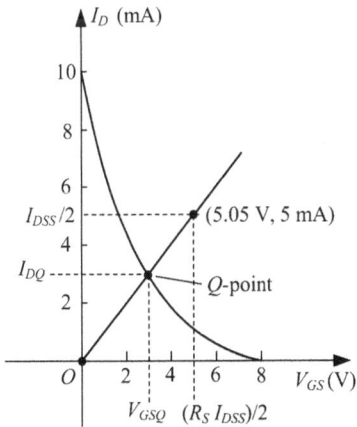

Fig. 4.40: Q-point obtained by graphical approach for Example 4.2.

Then in the output loop of Fig. 4.39, V_D can be determined at first as follows:

$$V_D = V_{DD} + I_D R_D$$
$$= -15\,V + (3.0\,mA \times 3\,k\Omega)$$
$$= -6.0\,V$$

and,

$$V_S = -I_S R_S$$
$$= -I_D R_S$$
$$= -(3.0\,mA \times 1.1\,k\Omega)$$
$$= -3.3\,V$$

So,

$$V_{DS} = V_D - V_S$$
$$= (-6.0\,V) - (-3.3\,V)$$
$$= -2.7\,V$$

For V_G, from the input loop in Fig. 4.39, the gate terminal is directly connected to the ground. So,

$$V_G = 0\,V$$

4.4.3 Common-gate configuration

This configuration is similar to the common-base configuration for the BJT amplifier in Section 3.3.6. Now, for simplicity, an example is used to show the analysis of the biasing circuit for the common-gate configuration. The results obtained are applicable to other common-gate configurations.

Example 4.3
The p-channel JFET common-gate configuration is shown in Fig. 4.41. Find the values of V_{GSQ}, I_{DQ}, V_{DS}, V_D, V_G and V_S for the network.

Solution
First, convert the JFET common-gate configuration to its DC equivalent circuit, as shown in Fig. 4.42. Note that C_1 and C_2 are removed.

The current through R_S is the source current I_S, and it is obvious that $I_S = I_D$. In the input loop, we obtain

$$V_{GS} = I_S R_S$$

Using the relationship of $I_S = I_D$,

$$V_{GS} = I_D R_S$$

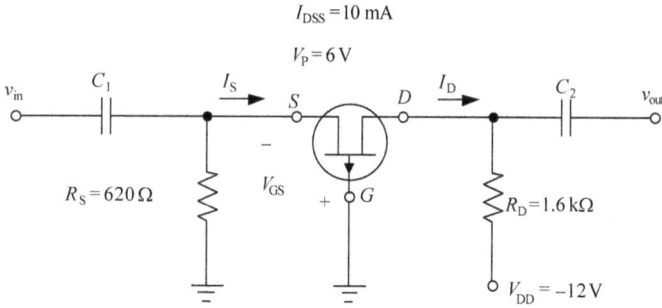

Fig. 4.41: Network of Example 4.3.

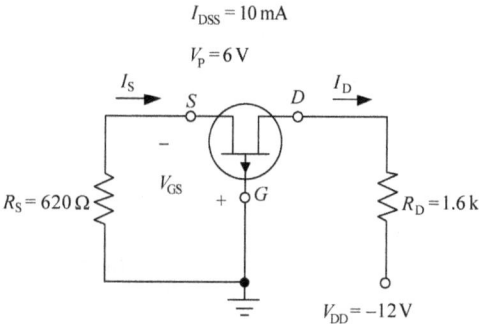

Fig. 4.42: DC equivalent circuit of Example 4.3.

So,

$$I_D = \frac{1}{R_S} V_{GS}$$

Thus, the $I_D \sim V_{GS}$ function defines a straight line through the origin. Then one more point is necessary to draw it. Any applicable values of V_{GS} can be used to determine the second point. So the point of $((R_S I_{DSS})/2, I_{DSS}/2)$, can be used. Substituting the values of I_{DSS} and R_S into the coordinates, the point of $(3.1\,V, 5\,mA)$ is obtained, as illustrated in Fig. 4.43.

The four points for the shorthand method to plot the transfer curve are summarized in Tab. 4.4.

Tab. 4.4: Four points to plot the transfer curve of Example 4.3.

Points	Values
$(0, I_{DSS})$	$(0\,V, 10\,mA)$
$(0.3\,V_P, I_{DSS}/2)$	$(1.8\,V, 5\,mA)$
$(0.5\,V_P, I_{DSS}/4)$	$(3\,V, 2.5\,mA)$
$(V_P, 0)$	$(6\,V, 0\,mA)$

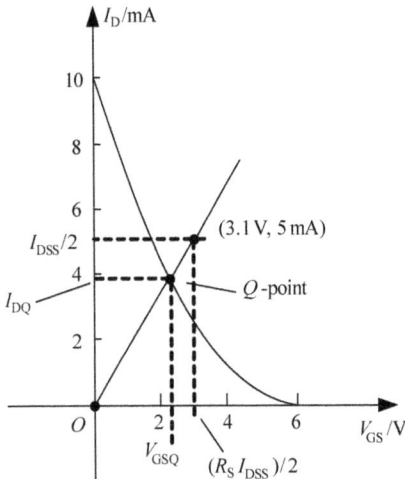

Fig. 4.43: *Q*-point obtained by the graphical approach for Example 4.3.

Then the transfer curve can be plotted, as illustrated in Fig. 4.43. So, the two relationships are ready, the transfer curve and the straight line. Then, the point where the two curves intersect is the solution to the DC analysis, the *Q*-point.

By drawing a horizontal line from the *Q*-point to the vertical I_D axis, I_{DQ} is determined. Further, by drawing a vertical line from the *Q*-point to the V_{GS} axis, V_{GSQ} is obtained, as illustrated in Fig. 4.43.

$$I_{DQ} \approx 3.8\,\text{mA}$$

$$V_{GSQ} \approx 2.4\,\text{V}$$

Then, in the output loop of Fig. 4.42, V_D can be determined at first as follows:

$$V_D = V_{DD} + I_D R_D$$
$$= -12\,\text{V} + (3.8\,\text{mA} \times 1.6\,\text{k}\Omega)$$
$$= -5.92\,\text{V}$$

and,

$$V_S = -I_S R_S$$
$$= -I_D R_S$$
$$= -(3.8\,\text{mA} \times 620\,\Omega)$$
$$= -2.36\,\text{V}$$

So,

$$V_{DS} = V_D - V_S$$
$$= (-5.92\,\text{V}) - (-2.36\,\text{V})$$
$$= -3.56\,\text{V}$$

For V_G, from input loop in Fig. 4.42, the gate terminal is directly connected to the ground. So,

$$V_G = 0\,V$$

4.4.4 Voltage-divider configuration

The voltage-divider bias configuration, applied to BJT amplifiers in Section 3.3.5, can also be applied to JFET amplifiers as illustrated in Fig. 4.44. Both share the exactly same basic construction, but the DC analysis of them is quite different. The nonzero magnitude of I_B for BJT amplifiers leads to equivalent and approximation methods. However, $I_G = 0\,A$ is a very convenient condition as the starting point for JFET amplifiers [3, 11, 12, 16].

The network of Fig. 4.44 is redrawn for the DC analysis, as shown in Fig. 4.45. Note that all the capacitors, C_1, C_2 and C_S, have been replaced by an "open-circuit" equivalent. Moreover, the voltage V_{DD} was separated into two equivalent voltages to separate the input and output loops. Since $I_G = 0\,A$, the currents satisfied that $I_{RG1} = I_{RG2}$. The voltage V_G, equal to the voltage across R_2, can be found using the voltage-divider rule as follows:

$$V_G = \frac{R_{G2}}{R_{G1} + R_{G2}} V_{DD}$$

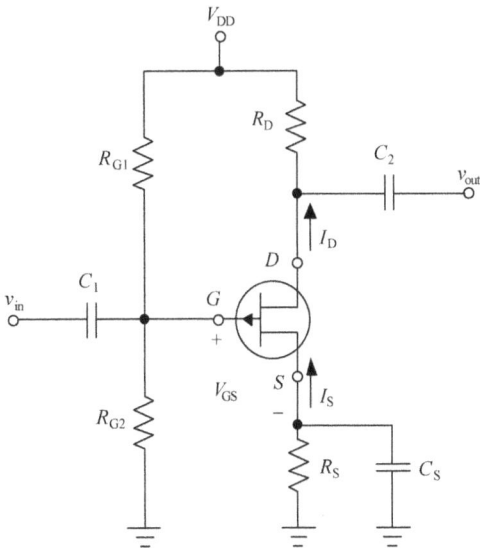

Fig. 4.44: Voltage-divider configuration for the p-channel JFET.

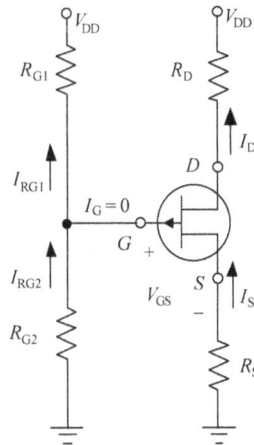

Fig. 4.45: Redrawn network of voltage-divider configuration for DC analysis.

Then, from the network in Fig. 4.45, we obtain

$$V_{GS} = V_G - V_S$$
$$= V_G - (-R_S I_S)$$
$$= V_G + R_S I_S$$
$$= V_G + R_S I_D$$

So,

$$I_D = \frac{V_{GS} - V_G}{R_S} \qquad (4.6)$$

It is obvious that the quantities V_G and R_S are fixed by the network construction. So the result is a function of $I_D \sim V_{GS}$, defining a straight line, although not through the origin.

The procedure for plotting this equation is the same as before, by letting either I_D or V_{GS} be zero to find two points:

$$I_D|_{V_{GS}=0} = -\frac{V_G}{R_S} \qquad (4.7)$$

and

$$V_{GS}|_{I_D=0} = V_G \qquad (4.8)$$

The two points defined above permit the drawing of a straight line to represent the function of $I_D \sim V_{GS}$. The intersection of the straight line with the transfer curve will define the operating point and the corresponding values of I_{DQ} and V_{GSQ}, as illustrated in Fig. 4.46.

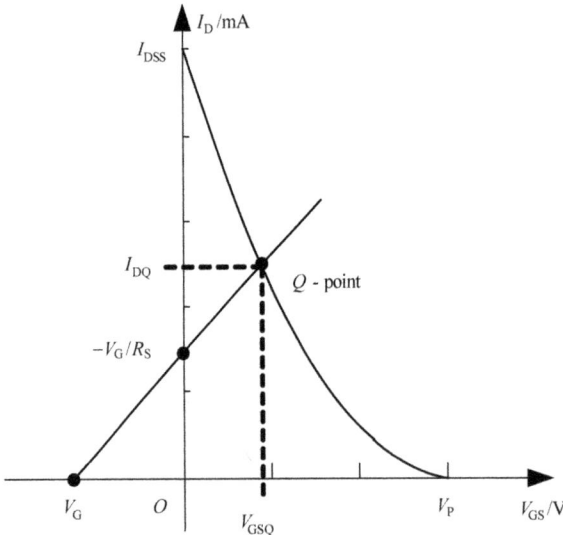

Fig. 4.46: Determining the Q-point of voltage-divider configuration.

Once the quiescent values of I_{DQ} and V_{GSQ} are determined, the remaining DC analysis can be carried out in the following manner:

$$V_{DS} = V_{DD} + I_D(R_D + R_S)$$
$$V_D = V_{DD} + I_D R_D$$
$$V_S = -I_D R_S$$
$$I_{RG1} = I_{RG2} = \frac{V_{DD}}{R_{G1} + R_{G2}}$$

Example 4.4

The p-channel JFET voltage-divider configuration is shown in Fig. 4.47. Find the values of V_{GSQ}, I_{DQ}, V_{DS}, V_{DG}, V_D and V_S for the network.

Solution

First, convert the JFET voltage-divider configuration to its DC equivalent circuit, as shown in Fig. 4.48. Note that C_1, C_2 and C_S are replaced with open circuits.

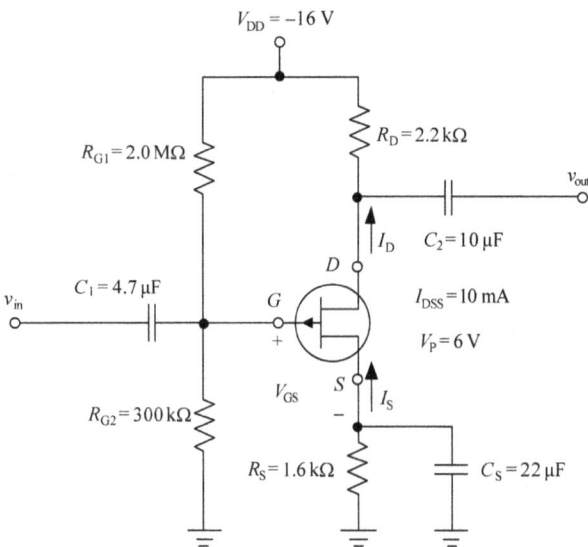

Fig. 4.47: Network of Example 4.4.

The transfer curve is obtained from the four points by the shorthand method, which is summarized in Tab. 4.5.

The transfer curve is located on the right-hand side of the vertical axis for the p-channel JFET, as illustrated in Fig. 4.49.

Tab. 4.5: Four points to plot the transfer curve of Example 4.4.

Points	Values
$(0, I_{DSS})$	$(0\,V, 10\,mA)$
$(0.3\,V_P, I_{DSS}/2)$	$(1.8\,V, 5\,mA)$
$(0.5\,V_P, I_{DSS}/4)$	$(3\,V, 2.5\,mA)$
$(V_P, 0)$	$(6\,V, 0\,mA)$

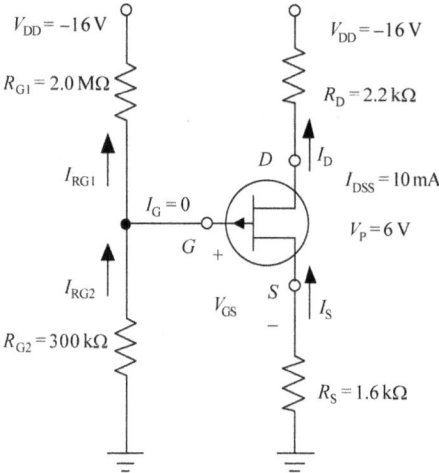

Fig. 4.48: DC equivalent circuit of Example 4.4.

Since $I_G = 0\,A$, the voltage V_G, can be found using the voltage-divider rule as follows:

$$V_G = \frac{R_{G2}}{R_{G1} + R_{G2}} V_{DD}$$

$$= \frac{300\,k\Omega}{2.0\,M\Omega + 300\,k\Omega} \times (-16\,V)$$

$$= -2.09\,V$$

Then it is time to plot the function of $I_D \sim V_{GS}$. By letting either I_D or V_{GS} be zero, find two points:

$$I_D|_{V_{GS}=0} = -\frac{V_G}{R_S}$$

$$= -\frac{-2.09\,V}{1.6\,k\Omega}$$

$$= 1.31\,mA$$

and

$$V_{GS}|_{I_D=0} = V_G = -2.09\,V$$

So the load line can be determined by the two points, $(0\,V, 1.31\,mA)$ and $(-2.09\,V, 0\,mA)$, as illustrated in Fig. 4.49.

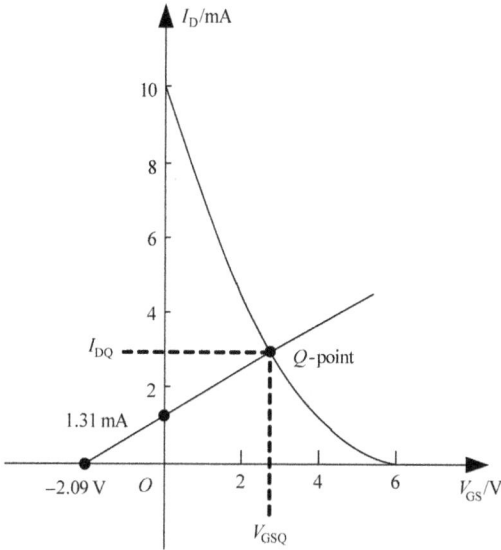

Fig. 4.49: Determining the Q-point of voltage-divider configuration.

Then the intersection of the load line with the transfer curve will define the Q-point. The corresponding values of I_{DQ} and V_{GSQ} read as follows:

$$I_{DQ} \approx 3.0\,\text{mA}$$

$$V_{GSQ} \approx 2.7\,\text{V}$$

Then,

$$V_D = V_{DD} + I_D R_D$$
$$= -16\,\text{V} + (3.0\,\text{mA} \times 2.2\,\text{k}\Omega)$$
$$= -9.4\,\text{V}$$

and,

$$V_S = -I_D R_S$$
$$= -(3.0\,\text{mA} \times 1.6\,\text{k}\Omega)$$
$$= -4.8\,\text{V}$$

So,

$$V_{DS} = V_D - V_S$$
$$= (-9.4\,\text{V}) - (-4.8\,\text{V})$$
$$= -4.6\,\text{V}$$

or from the network, we obtain

$$V_{DS} = V_{DD} + I_D(R_D + R_S)$$
$$= -16\,\text{V} + 3.0\,\text{mA} \times (2.2\,\text{k}\Omega + 1.6\,\text{k}\Omega)$$
$$= -4.6\,\text{V}$$

Finally,

$$V_{DG} = V_D - V_G$$
$$= (-9.4\,\text{V}) - (-2.09\,\text{V})$$
$$= -7.31\,\text{V}$$

or,

$$V_{DG} = V_{DS} + V_{SG}$$
$$= V_{DS} - V_{GSQ}$$
$$\approx (-4.6\,\text{V}) - (2.7\,\text{V})$$
$$= -7.3\,\text{V}$$

Note that, in Fig. 4.46, the load line has an intersection with the vertical axis, which is $-V_G/R_S$, and the value of V_G is fixed and determined by two resistors, R_{G1} and R_{G2}. So, increasing R_S will reduce the values of $-V_G/R_S$ and thus cause the intersection to move down vertically and the Q-point to move along the transfer curve, as indicated by Fig. 4.50.

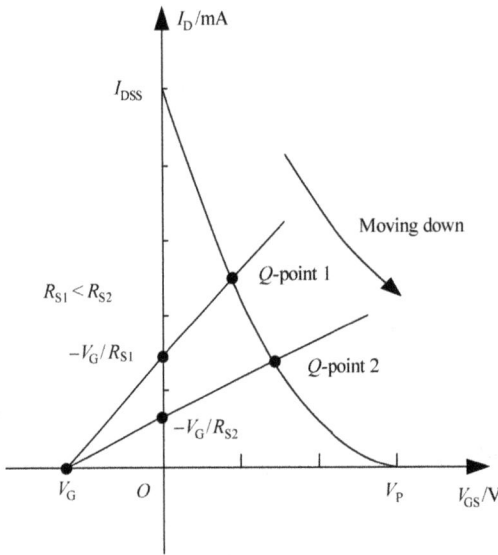

Fig. 4.50: Influence of R_S on the Q-point for voltage-divider configuration.

4.5 MOSFET DC biasing configurations

The DC analysis of biasing configurations for MOSFETs is partially the same as those for JFETs. However, more importantly, there are differences in characteristics between MOSFETs and JFETs, and these should be involved in the procedure of DC analysis.

4.5.1 Depletion-type MOSFETs

Most parts of the transfer curves of JFETs and depletion-type MOSFETs are similar, which means that the DC analysis of them is almost the same. The main difference between the two is that V_{GS} of depletion-type MOSFETs can be negative for the p-channel and positive for the n-channel, and I_D greater than I_{DSS}.

For simplicity, several examples are used to show the analysis of depletion-type MOSFET biasing circuits. The general ideas can be applied to other depletion-type MOSFET biasing configurations.

Example 4.5
The p-channel depletion-type MOSFET voltage-divider configuration is shown in Fig. 4.51.
(1) Find the values of V_{GSQ}, I_{DQ} and V_{DS} for the network.
(2) Do the same work when $R_S = 160\,\Omega$ [1, 11–13].

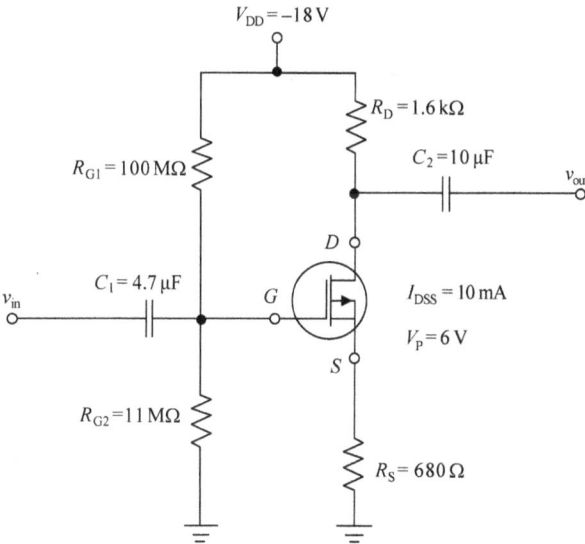

Fig. 4.51: Network of Example 4.5.

Solution

(1) First, convert the depletion-type MOSFET voltage-divider configuration to its DC equivalent circuit, as shown in Fig. 4.52. Note that C_1 and C_2 are replaced with open circuits.

On the one hand, the main part of the transfer characteristics are obtained in the same way as for JFETs, from the four points by the shorthand method (summarized in Tab. 4.6).

On the other hand, due to the fact that Shockley's equation defines the curve that increases more rapidly as V_{GS} becomes more negative, one more point will be defined at $V_{GS} = -1$ V. Substituting it into Shockley's equation yields

$$I_D = I_{DSS}\left(1 - \frac{V_{GS}}{V_P}\right)^2$$

$$= 10\,\text{mA}\left(1 - \frac{-1\,\text{V}}{6\,\text{V}}\right)^2$$

$$= 13.6\,\text{mA}$$

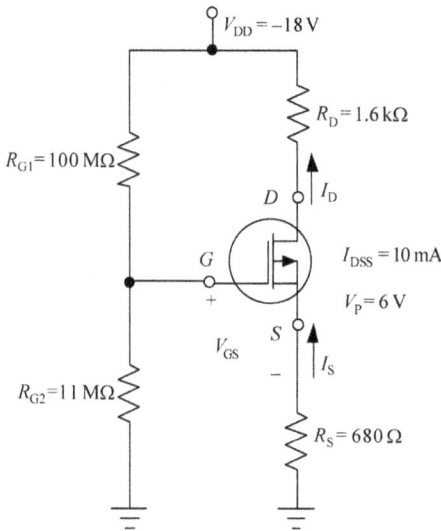

Fig. 4.52: DC equivalent circuit of Example 4.5.

Tab. 4.6: Four points to plot transfer curve of Example 4.5.

Points	Values
$(0, I_{DSS})$	$(0\,\text{V}, 10\,\text{mA})$
$(0.3\,V_P, I_{DSS}/2)$	$(1.8\,\text{V}, 5\,\text{mA})$
$(0.5\,V_P, I_{DSS}/4)$	$(3\,\text{V}, 2.5\,\text{mA})$
$(V_P, 0)$	$(6\,\text{V}, 0\,\text{mA})$

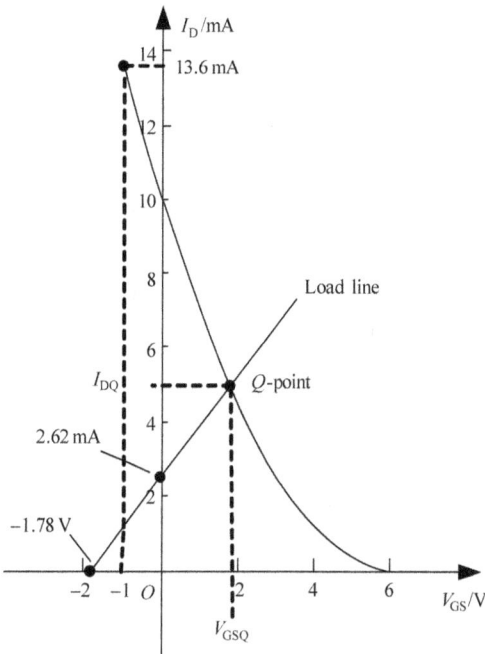

Fig. 4.53: Determining the Q-point of depletion-type MOSFET voltage-divider configuration.

So, the fifth point for the transfer curve is $(-1\,V, 13.6\,mA)$. The main part of the transfer curve its location on the right-hand side of the vertical axis for the p-channel, as illustrated in Fig. 4.53.

Then, for the load line, in the input loop the starting point is $I_G = 0\,A$. The voltage V_G can be found using the voltage-divider rule as follows:

$$V_G = \frac{R_{G2}}{R_{G1} + R_{G2}} V_{DD}$$
$$= \frac{11\,M\Omega}{100\,M\Omega + 11\,M\Omega} \times (-18\,V)$$
$$= -1.78\,V$$

Then it is time to plot the load line of $I_D \sim V_{GS}$ from Eq. (4.6), defining a straight line, though not through the origin. By letting either variable be zero, find two points:

$$I_D|_{V_{GS}=0} = -\frac{V_G}{R_S}$$
$$= -\frac{-1.78\,V}{680\,\Omega}$$
$$= 2.62\,mA$$

and

$$V_{GS}|_{I_D=0} = V_G$$
$$= -1.78\,V$$

So the load line can be determined by the two points $(0\,V, 2.62\,mA)$ and $(-1.78\,V, 0\,mA)$ as illustrated in Fig. 4.53.

Then the intersection of the load line with the transfer curve will define the Q-point. The corresponding values of I_{DQ} and V_{GSQ} read as follows:

$$I_{DQ} \approx 5.1\,mA$$
$$V_{GSQ} \approx 1.8\,V$$

Then, from Fig. 4.52, we obtain

$$V_D = V_{DD} + I_D R_D$$
$$= -18\,V + (5.1\,mA \times 1.6\,k\Omega)$$
$$= -9.84\,V$$

and,

$$V_S = -I_D R_S$$
$$= -(5.1\,mA \times 680\,\Omega)$$
$$= -3.47\,V$$

So,

$$V_{DS} = V_D - V_S$$
$$= (-9.84\,V) - (-3.47\,V)$$
$$= -6.37\,V$$

or from the output loop in the network, we obtain

$$V_{DS} = V_{DD} + I_D(R_D + R_S)$$
$$= -18\,V + 5.1\,mA \times (1.6\,k\Omega + 680\,\Omega)$$
$$= -6.37\,V$$

(2) When $R_S = 160\,\Omega$, the DC equivalent circuit is the same as in Fig. 4.52, except for the value of R_S; the transfer characters are same as in Fig. 4.53.

The voltage V_G found by the voltage-divider rule is the same:

$$V_G = \frac{R_{G2}}{R_{G1} + R_{G2}} V_{DD}$$
$$= -1.78\,V$$

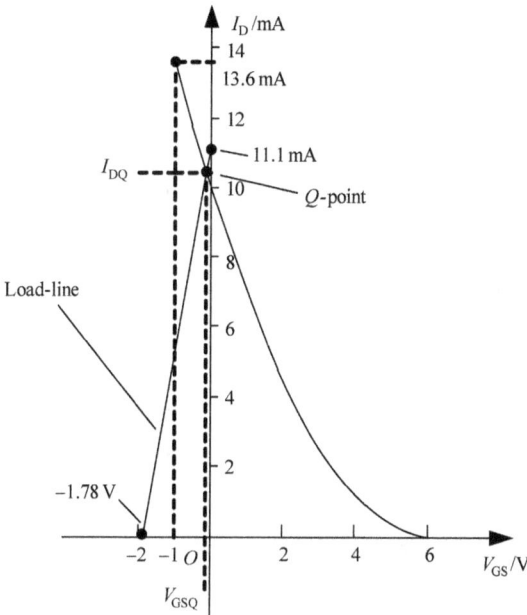

Fig. 4.54: Determining the Q-point when $R_S = 160\,\Omega$.

However, the load line of $I_D \sim V_{GS}$ as defined by Eq. (4.6), will definitely change. By letting either I_D or V_{GS} be zero, find two points:

$$I_D|_{V_{GS}=0} = -\frac{V_G}{R_S}$$
$$= -\frac{-1.78\,\text{V}}{160\,\Omega}$$
$$= 11.1\,\text{mA}$$

and

$$V_{GS}|_{I_D=0} = V_G$$
$$= -1.78\,\text{V}$$

So the load line can be determined by the two points, $(0\,\text{V}, 11.1\,\text{mA})$ and $(-1.78\,\text{V}, 0\,\text{mA})$, as illustrated in Fig. 4.54.

Then the intersection of the load line with the transfer curve will define the Q-point. The corresponding values of I_{DQ} and V_{GSQ} read as follows:

$$I_{DQ} \approx 10.5\,\text{mA}$$
$$V_{GSQ} \approx -0.15\,\text{V}$$

From Fig. 4.52 with $R_S = 160\,\Omega$, we obtain

$$V_D = V_{DD} + I_D R_D$$
$$= -18\,V + (10.5\,mA \times 1.6\,k\Omega)$$
$$= -1.2\,V$$

and,

$$V_S = -I_D R_S$$
$$= -(10.5\,mA \times 160\,\Omega)$$
$$= -1.68\,V$$

So,

$$V_{DS} = V_D - V_S$$
$$= (-1.2\,V) - (-1.68\,V)$$
$$= 0.48\,V$$

or from the output loop in the network, we obtain

$$V_{DS} = V_{DD} + I_D(R_D + R_S)$$
$$= -18\,V + 10.5\,mA \times (1.6\,k\Omega + 160\,\Omega)$$
$$= 0.48\,V$$

Example 4.6
The p-channel depletion-type MOSFET self-bias configuration is shown in Fig. 4.55. Find the values of V_{GSQ}, I_{DQ} and V_D for the network.

Solution
First, convert the depletion-type MOSFET self-bias configuration to its DC equivalent circuit, as shown in Fig. 4.56. Note that C_1 and C_2 are removed due to their high impedance for the DC equivalent, and R_G is removed due to $I_G = 0\,A$.

On the one hand, the main part of transfer characteristics are obtained in the same way as for JFETs, from the four points by the shorthand method (summarized in Tab. 4.7).

Tab. 4.7: Four points to plot the transfer curve of Example 4.6.

Points	Values
$(0, I_{DSS})$	$(0\,V, 10\,mA)$
$(0.3\,V_P, I_{DSS}/2)$	$(1.8\,V, 5\,mA)$
$(0.5\,V_P, I_{DSS}/4)$	$(3\,V, 2.5\,mA)$
$(V_P, 0)$	$(6\,V, 0\,mA)$

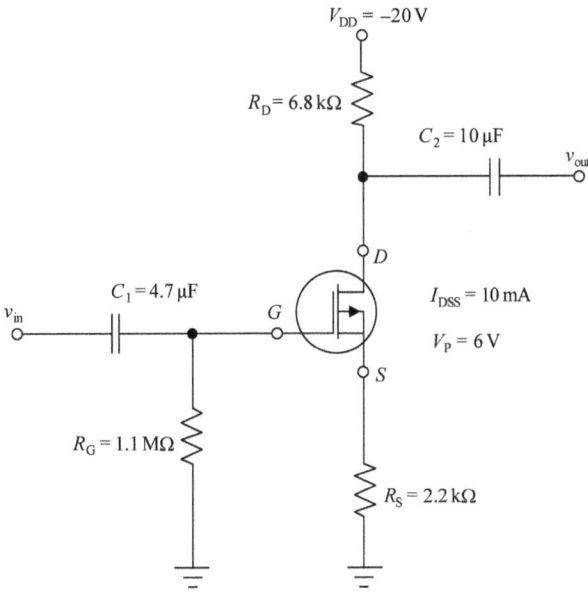

Fig. 4.55: Network of Example 4.6.

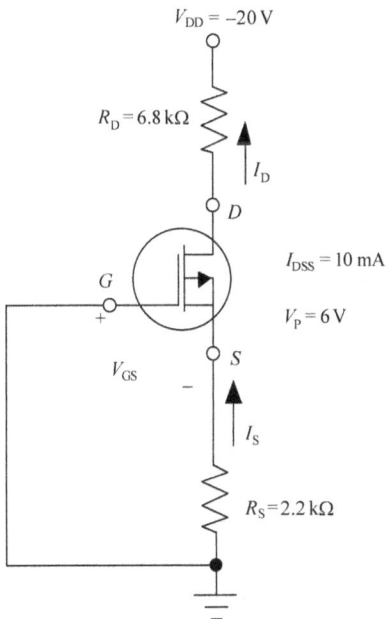

Fig. 4.56: DC equivalent circuit of Example 4.6.

On the other hand, due to the fact that Shockley's equation defines a curve that increases more rapidly as V_{GS} becomes more negative, one more point will be defined at $V_{GS} = -1\,\text{V}$. Substituting it into Shockley's equation yields

$$I_D = I_{DSS}\left(1 - \frac{V_{GS}}{V_P}\right)^2$$

$$= 10\,\text{mA}\left(1 - \frac{-1\,\text{V}}{6\,\text{V}}\right)^2$$

$$= 13.6\,\text{mA}$$

So, the fifth point for the transfer curve is $(-1\,\text{V}, 13.6\,\text{mA})$. The main part of the transfer curve is located on the right-hand side of the vertical axis for the p-channel, as illustrated in Fig. 4.57.

Now it is time to determine the relationship of $I_D \sim V_{GS}$. From Fig. 4.56, it is obvious that $I_S = I_D$. In the input loop, we obtain

$$V_{GS} = I_S R_S$$

Using the relationship of $I_S = I_D$,

$$V_{GS} = I_D R_S$$

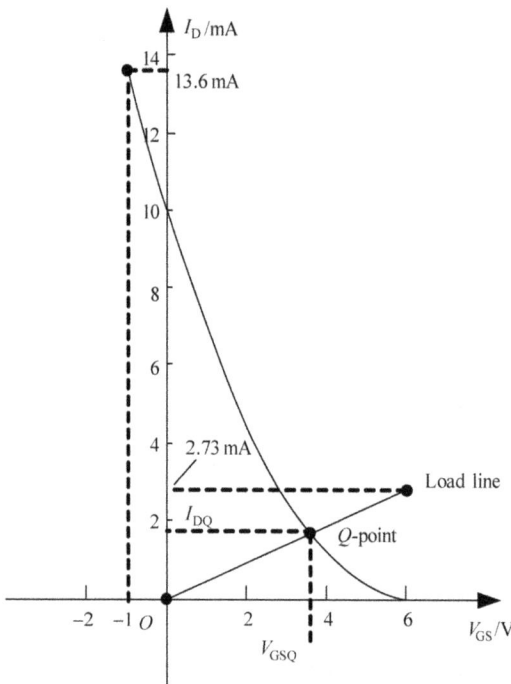

Fig. 4.57: Determining the Q-point of depletion-type MOSFET self-bias configuration.

So,

$$I_D = \frac{1}{R_S} V_{GS}$$

Note that it is a straight line through the origin. Then one more point is necessary to draw it. Substituting the values of $V_{GS} = 6\,V$, we obtain

$$I_D = \frac{6\,V}{2.2\,k\Omega} = 2.73\,mA$$

So the point is $(6\,V, 2.73\,mA)$ is obtained and the straight line can be determined, as illustrated in Fig. 4.57.

Then the intersection of the load line with the transfer curve will define the Q-point. The corresponding values of I_{DQ} and V_{GSQ} read as follows:

$$I_{DQ} \approx 1.8\,mA$$

$$V_{GSQ} \approx 3.7\,V$$

Then, from Fig. 4.56, we obtain

$$V_D = V_{DD} + I_D R_D$$
$$= -20\,V + (1.8\,mA \times 6.8\,k\Omega)$$
$$= -7.76\,V$$

Example 4.7
The p-channel depletion-type MOSFET fixed-bias configuration is shown in Fig. 4.58. Find the values of V_{DS} for the network.

Solution
This configuration looks like the fixed-bias configuration for the JFET, shown in Fig. 4.27, except that V_{GG} exists. This example could be a special case of fixed bias, and its DC equivalent circuit is quite simple, as shown in Fig. 4.59.

From the DC equivalent circuit of Fig. 4.59, we obtain

$$V_{GS} = 0\,V$$

and the drain current I_D is always determined by Shockley's equation. So,

$$I_D = I_{DSS} \left(1 - \frac{V_{GS}}{V_P} \right)^2$$
$$= I_{DSS}$$
$$= 10\,mA$$

$V_{DD} = -20\,V$

$R_D = 1.6\,k\Omega$

$C_2 = 10\,\mu F$

v_{out}

D

$C_1 = 4.7\,\mu F$

v_{in}

G

$I_{DSS} = 10\,mA$

$V_P = 4\,V$

S

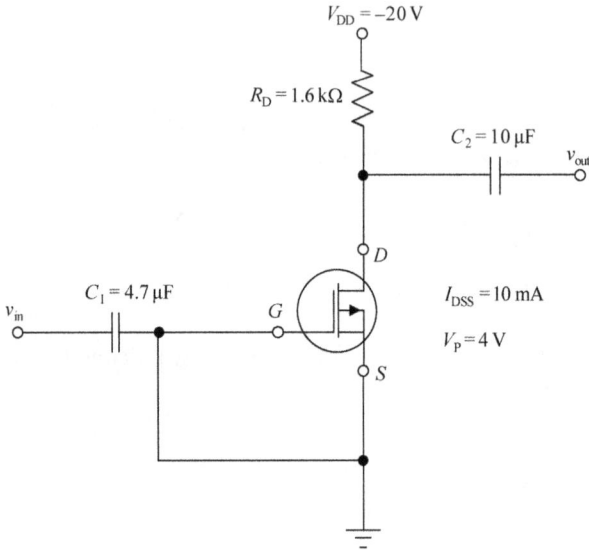

Fig. 4.58: Network of Example 4.7.

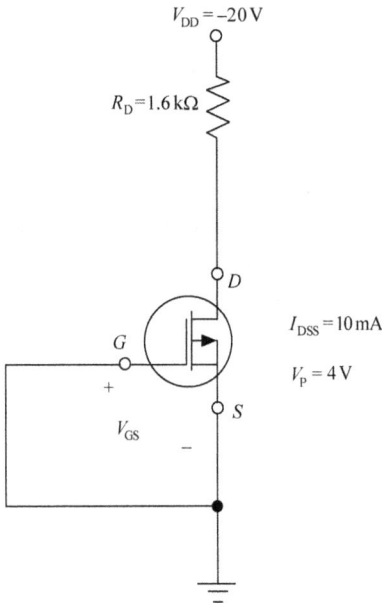

$V_{DD} = -20\,V$

$R_D = 1.6\,k\Omega$

D

G

$I_{DSS} = 10\,mA$

$V_P = 4\,V$

$+$

S

V_{GS}

$-$

Fig. 4.59: DC equivalent circuit of Example 4.7.

Then from Fig. 4.59, we obtain

$$V_{DS} = V_D$$
$$= V_{DD} + I_D R_D$$
$$= -20\,V + (10\,mA \times 1.6\,k\Omega)$$
$$= -4.0\,V$$

and it can be seen that the transfer curve is not used in this special case.

4.5.2 Enhancement-type MOSFETs

The transfer characteristics of the enhancement-type MOSFET are quite different from those for the JFET or depletion-type MOSFETs; I_D only exists when V_{GS} is more negative than $V_{GS(Th)}$ for the p-channel devices, as shown in Fig. 4.23, or more positive than $V_{GS(Th)}$ for the n-channel devices, as shown in Fig. 4.25. When I_D is nonzero, it is defined by Eq. (4.2), as follows:

$$I_D = k\left(V_{GS} - V_{GS(Th)}\right)^2$$

where k is a constant, determined by Eq. (4.3), as follows:

$$k = \frac{I_{D(on)}}{\left(V_{GS(on)} - V_{GS(Th)}\right)^2}$$

where $I_{D(on)}$ and $V_{GS(on)}$ are the parameters at a particular point on the device characteristics.

To be straightforward, several examples are introduced to show the analysis of enhancement-type MOSFET biasing circuits. The general ideas are applicable to other enhancement-type MOSFET biasing configurations.

Example 4.8

The p-channel enhancement-type MOSFET voltage-divider configuration is shown in Fig. 4.60. Find the values of V_{GSQ}, I_{DQ} and V_{DS}.

Solution

Convert the enhancement-type MOSFET voltage-divider configuration to its DC equivalent circuit, as shown in Fig. 4.61. Note that C_1 and C_2 are replaced with open circuits. Then for the load line, in the input loop the starting point is $I_G = 0\,A$. The voltage V_G can be found using the voltage-divider rule as follows:

$$V_G = \frac{R_{G2}}{R_{G1} + R_{G2}} V_{DD}$$
$$= \frac{16\,M\Omega}{20\,M\Omega + 16\,M\Omega} \times (-40\,V)$$
$$= -17.78\,V$$

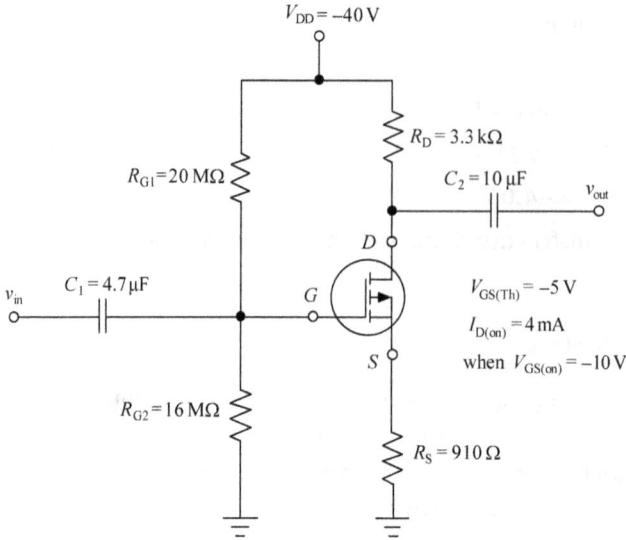

Fig. 4.60: Network of Example 4.8.

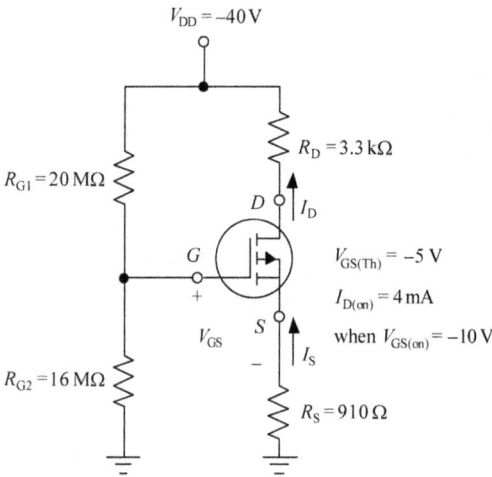

Fig. 4.61: DC equivalent circuit of Example 4.8.

Now it is time to plot the load line of $I_D \sim V_{GS}$ from Eq. (4.6), defining a straight line, although not through the origin. By letting either variable be zero, find two points:

$$I_D|_{V_{GS}=0} = -\frac{V_G}{R_S}$$
$$= -\frac{-17.78\,V}{910\,\Omega}$$
$$= 19.54\,mA$$

and

$$V_{GS}|_{I_D=0} = V_G = -17.78\,V$$

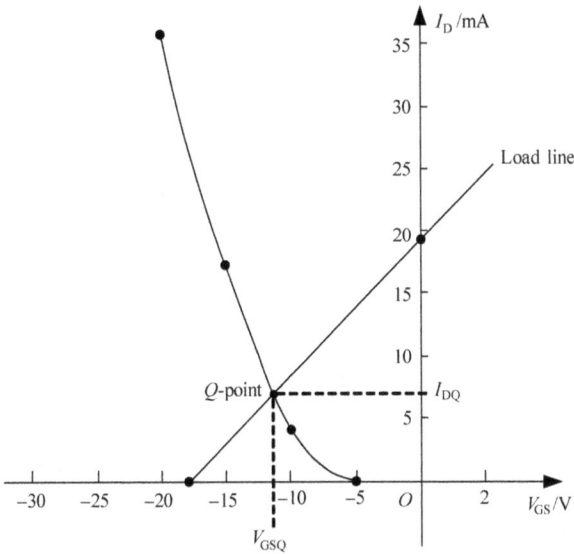

Fig. 4.62: Determining Q-point of enhancement-type MOSFET voltage-divider configuration.

So the load line can be determined by the two points, $(0\,V, 19.54\,mA)$ and $(-17.78\,V, 0\,mA)$, as illustrated in Fig. 4.62.

Then next step is to plot transfer characteristics by the shorthand method. As it is given that $I_{D(on)} = 4\,mA$ when $V_{GS(on)} = -10\,V$ and $V_{GS(Th)} = -5\,V$, the constant k is determined by Eq. (4.3) as follows:

$$k = \frac{I_{D(on)}}{\left(V_{GS(on)} - V_{GS(Th)}\right)^2}$$
$$= \frac{4\,mA}{(-10\,V - (-5\,V))^2}$$
$$= 0.16 \times 10^{-3}\,A/V^2$$

So, when V_{GS} is more negative than $V_{GS(Th)}$ for the p-channel devices, I_D is nonzero and defined by Eq. (4.2), as follows:

$$I_D = k\left(V_{GS} - V_{GS(Th)}\right)^2$$
$$= 0.16 \times 10^{-3}\left(V_{GS} - (-5\,V)\right)^2$$

and two more points are necessary to sketch the transfer curve. They are

$$I_D|_{V_{GS}=-15} = k\left(V_{GS} - V_{GS(Th)}\right)^2$$
$$= 0.16 \times 10^{-3} \times (-15 - (-5\,V))^2$$
$$= 16\,mA$$

Tab. 4.8: Four points to plot the transfer curve of Example 4.8.

Points: (V_{GS}, I_D)	Values
($V_{GS(Th)}$, I_D)	(−5 V, 0 mA)
($V_{GS(on)}$, $I_{D(on)}$)	(−10 V, 4 mA)
(V_{GS1}, I_{D1})	(−15 V, 16 mA)
(V_{GS2}, I_{D2})	(−20 V, 36 mA)

and

$$I_D|_{V_{GS}=-20} = k \left(V_{GS} - V_{GS(Th)}\right)^2$$
$$= 0.16 \times 10^{-3} \times (-20 - (-5\,\text{V}))^2$$
$$= 36\,\text{mA}$$

The four points to plot transfer characteristics by the shorthand method are summarized in Tab. 4.8.

So, from the four points, the transfer curve is obtained. It is located on the left-hand side of the vertical axis for p-channel devices, as illustrated in Fig. 4.62. Then the intersection of the load line with the transfer curve will define the Q-point. The corresponding values of I_{DQ} and V_{GSQ} read as follows:

$$I_{DQ} \approx 6.1\,\text{mA}$$
$$V_{GSQ} \approx -11.2\,\text{V}$$

Then, from Fig. 4.61, we obtain

$$V_D = V_{DD} + I_D R_D$$
$$= -40\,\text{V} + (6.1\,\text{mA} \times 3.3\,\text{k}\Omega)$$
$$= -19.87\,\text{V}$$

and,

$$V_S = -I_D R_S$$
$$= -(6.1\,\text{mA} \times 910\,\Omega)$$
$$= -5.55\,\text{V}$$

So,

$$V_{DS} = V_D - V_S$$
$$= (-19.87\,\text{V}) - (-5.55\,\text{V})$$
$$= -14.32\,\text{V}$$

or from output loop in the network, we obtain

$$V_{DS} = V_{DD} + I_D(R_D + R_S)$$
$$= -40\,\text{V} + 6.1\,\text{mA} \times (3.3\,\text{k}\Omega + 910\,\Omega)$$
$$= -14.32\,\text{V}$$

Example 4.9

The p-channel enhancement-type MOSFET drain-feedback bias configuration is shown in Fig. 4.63. Find the values of V_{GSQ} and I_{DQ}.

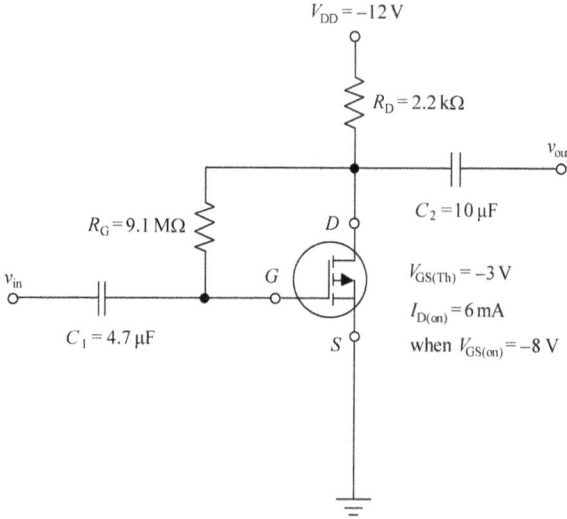

Fig. 4.63: Network of Example 4.9.

Solution

The resistor R_G feeds a sufficiently large voltage from the terminal D to the terminal G, to drive the p-channel enhancement-type MOSFET in the "ON" state. Here, I_G is zero and the voltage drop across R_G is also zero; the DC equivalent circuit is shown in Fig. 4.64. Note that C_1 and C_2 are replaced with open circuits.

Then from the output loop in Fig. 4.64, we obtain

$$V_{DS} = V_{DD} + I_D R_D$$

In the input loop, it is obvious that

$$V_D = V_G$$

So,

$$V_{GS} = V_{DD} + I_D R_D$$

or in the form of $I_D \sim V_{GS}$, it is

$$I_D = \frac{V_{GS} - V_{DD}}{R_D}$$

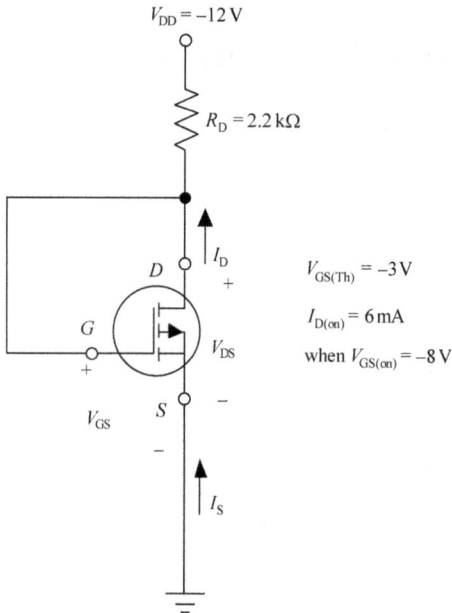

$V_{DD} = -12\,V$

$R_D = 2.2\,k\Omega$

$V_{GS(Th)} = -3\,V$

$I_{D(on)} = 6\,mA$

when $V_{GS(on)} = -8\,V$

Fig. 4.64: DC equivalent circuit of Example 4.9.

The procedure for plotting this straight line is the same as before, i.e., by letting either variable be zero to find two points:

$$I_D|_{V_{GS}=0} = -\frac{V_{DD}}{R_D}$$
$$= -\frac{-12\,V}{2.2\,k\Omega}$$
$$= 5.45\,mA$$

and

$$V_{GS}|_{I_D=0} = V_{DD}$$
$$= -12\,V$$

So the load line can be determined by the two points (0 V, 5.45 mA) and (−12 V, 0 mA), as illustrated in Fig. 4.65.

Then next step is to plot transfer characteristics by the shorthand method. As it is given that $I_{D(on)} = 6\,mA$ when $V_{GS(on)} = -8\,V$ and $V_{GS(Th)} = -3\,V$, the constant k is determined by Eq. (4.3) as follows:

$$k = \frac{I_{D(on)}}{\left(V_{GS(on)} - V_{GS(Th)}\right)^2}$$
$$= \frac{6\,mA}{(-8\,V - (-3\,V))^2}$$
$$= 0.24 \times 10^{-3}\,A/V^2$$

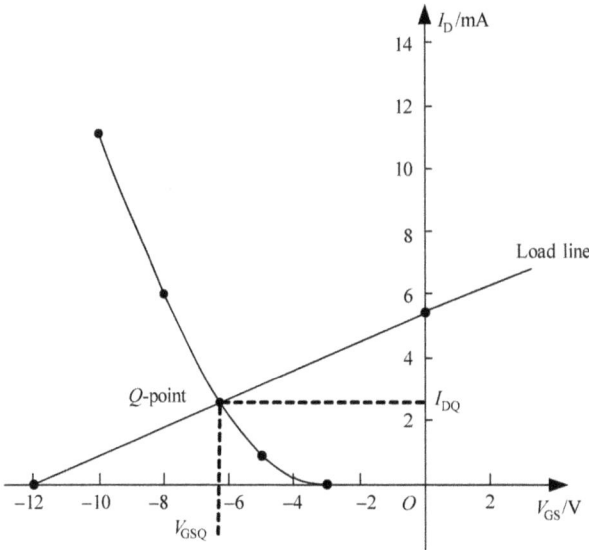

Fig. 4.65: Determining the Q-point of enhancement-type MOSFET feedback-bias configuration.

So, when V_{GS} is more negative than $V_{GS(Th)}$ for p-channel devices, I_D is nonzero and defined by Eq. (4.2), as follows:

$$I_D = k\left(V_{GS} - V_{GS(Th)}\right)^2$$
$$= 0.24 \times 10^{-3}\left(V_{GS} - (-3\text{ V})\right)^2$$

Two more points are necessary to sketch the transfer curve:

$$I_D|_{V_{GS}=-5} = k\left(V_{GS} - V_{GS(Th)}\right)^2$$
$$= 0.24 \times 10^{-3} \times (-5 - (-3\text{ V}))^2$$
$$= 0.96\text{ mA}$$

and

$$I_D|_{V_{GS}=-10} = k\left(V_{GS} - V_{GS(Th)}\right)^2$$
$$= 0.24 \times 10^{-3} \times (-10 - (-3\text{ V}))^2$$
$$= 11.76\text{ mA}$$

The four points to plot the transfer characteristics by the shorthand method are summarized in Tab. 4.9.

So, from the four points, the transfer curve is obtained, located on the left-hand side of the vertical axis for p-channel devices, as illustrated in Fig. 4.65.

Tab. 4.9: Four points to plot the transfer curve of Example 4.9.

Points: (V_{GS}, I_D)	Values
($V_{GS(Th)}$, I_D)	(-3 V, 0 mA)
(V_{GS1}, I_{D1})	(-5 V, 0.96 mA)
($V_{GS(on)}$, $I_{D(on)}$)	(-8 V, 6 mA)
(V_{GS2}, I_{D2})	(-10 V, 11.76 mA)

Then the intersection of the load line with the transfer curve will define the Q-point. The corresponding values of I_{DQ} and V_{GSQ} read as follows:

$$I_{DQ} \approx 2.6 \, \text{mA}$$
$$V_{GSQ} \approx -6.4 \, \text{V}$$

Moreover, for this configuration, $V_{GSQ} = V_{DSQ}$.

4.6 FET AC analysis

As discussed in Chapter 3, the BJT device controls a large output (collector) current by means of a relatively small input (base) current. In this section, FET AC analysis will be carried out to show that the FET device controls an output (drain) current by means of a small input (gate) voltage. Generally, the BJT is a current-controlled device and the FET is a voltage-controlled device. For both of them, the output current is the controlled variable [1, 3, 11–13, 16].

FET amplifiers provide satisfactory voltage gain with the additional feature of high input impedance and can be used as linear amplifiers or as digital logic circuits (this will not be part of the discussion of this textbook). JFET and depletion-type MOSFET amplifiers have similar voltage gains. However, the enhancement-type MOSFET is widely-used in low power-consumption digital circuitry.

For FET amplifier, due to the very high input impedance, the input current is generally assumed to be zero and the current gain is an undefined quantity. The voltage gain of an FET amplifier is generally less than that of the BJT amplifier. However, the FET amplifier provides much higher input impedance than that of a BJT configuration. Output impedance values are comparable for both BJT and FET circuits [3, 11, 16].

Whereas the BJT has an amplification factor of β, the FET has a transconductance factor, denoted as g_m. The AC equivalent model is somewhat simpler than that of the BJT.

4.6.1 Transconductance

The FET AC analysis is a process that requires an equivalent circuit called the small-signal AC mode, which reflects that AC voltage V_{GS} controls the AC level of the current I_D.

The relationship between I_D and V_{GS} is defined by Shockley's equation [3, 11, 16], Eq. (4.1), as follows:

$$I_D = I_{DSS} \left(1 - \frac{V_{GS}}{V_P} \right)^2$$

Then the change in the drain current I_D that results from a change in V_{GS} can be determined using the transconductance factor g_m in the following manner:

$$\Delta I_D = g_m \Delta V_{GS} \qquad (4.9)$$

The prefix "trans" used in "transconductance" means that it establishes a relationship between an output and an input quantity. The root of "conductance" reveals that g_m is a voltage-to-current ratio, which has a similar physical meaning to the conductance of a resistor, that is, $G = 1/R = I/V$ with the unit of Siemens. Moreover, it is obvious that [1, 3, 11–13, 16]

$$g_m = \frac{\Delta I_D}{\Delta V_{GS}} \qquad (4.10)$$

In fact, it is obvious that g_m is the slope of the transfer characteristics at the operation point, as described analytically by Eq. (4.10) and graphically in Fig. 4.66. So, g_m can be

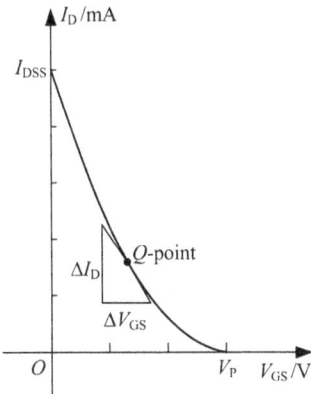

Fig. 4.66: Definition of g_m by transfer characteristics.

obtained at any Q-point on the characteristics by simply choosing a finite increment in V_{GS} about the Q-point and the corresponding change in I_D and substituting them into Eq. (4.10).

The above graphical approach is limited by the accuracy of the transfer characteristics plot and the manual choice of increment. Also note that from the monotonicity of the transfer curve, g_m will be negative for p-channel devices and positive for n-channel devices.

However, a more accurate method of determining g_m is a mathematical one, that is, finding directly the derivative of the function $I_D \sim V_{GS}$. So,

$$
\begin{aligned}
g_m &= \frac{dI_D}{dV_{GS}}\Big|_{Q\text{-point}} \\
&= \frac{d}{dV_{GS}}\left[I_{DSS}\left(1 - \frac{V_{GS}}{V_P}\right)^2\right] \\
&= I_{DSS}\frac{d}{dV_{GS}}\left[\left(1 - \frac{V_{GS}}{V_P}\right)^2\right] \\
&= 2I_{DSS}\left(1 - \frac{V_{GS}}{V_P}\right)\frac{d}{dV_{GS}}\left[\left(1 - \frac{V_{GS}}{V_P}\right)\right] \\
&= 2I_{DSS}\left(1 - \frac{V_{GS}}{V_P}\right)\left(-\frac{1}{V_P}\right) \\
&= \frac{2I_{DSS}}{V_P^2}(V_{GS} - V_P)
\end{aligned}
$$

So that

$$
g_m = \frac{2I_{DSS}}{V_P^2}(V_{GS} - V_P) \tag{4.11}
$$

Note that for both p-channel and n-channel devices, $|V_{GS}| < |V_P|$, so g_m will still be negative for p-channel devices and positive for n-channel devices. This is the same as the conclusion drawn from the graphical method. Also note that g_m is not a constant along the transfer curve.

4.6.2 Input impedance

The input impedance Z_i of commonly used FET is so large that is assumed as an open circuit. In equation form, it is [1, 3, 11]

$$
Z_i = \infty \, \Omega
$$

Z_i of a practical JFET has a typical value of $10^9 \, \Omega$, and MOSFET has $10^{12} \, \Omega$ to $10^{15} \, \Omega$, typically.

4.6.3 Output impedance

The output impedance Z_o of the FET is similar in magnitude to that of conventional the BJT. It is defined on the drain characteristics of Fig. 4.67 as the slope of the characteristic curve at the Q-point. The more horizontal the curve, the greater the output impedance. If completely horizontal, the output impedance Z_o is infinite, or an open circuit, which is an ideal situation [1, 6].

In equation form, it is

$$Z_o = r_d$$
$$= \frac{\Delta V_{GS}}{\Delta I_D}\Big|_{Q\text{-point}} \tag{4.12}$$

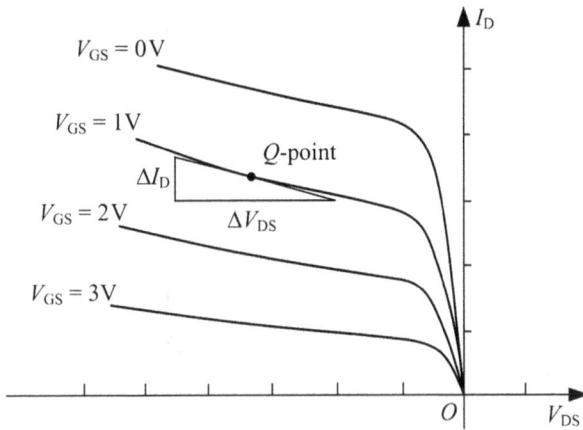

Fig. 4.67: Definition of r_d by drain characteristics.

4.6.4 FET AC equivalent circuit

After the introduction of the important parameters of the AC equivalent circuit, transconductance g_m, input impedance Z_i and output impedance Z_o, a model for the FET in the AC domain can be set up. The control of I_d by V_{gs} is included as a controlled current source $g_m V_{gs}$ connected between D and S, as shown in Fig. 4.68. The arrow of the current source points from D to S or from S to D to show the actual current operation for practical FET types.

As illustrated in Fig. 4.68, the input impedance Z_i is represented by the open circuit at the input loop. The output impedance Z_o is represented by the resistor r_d from terminal D to terminal S. Note that the gate-to-source voltage is now represented by

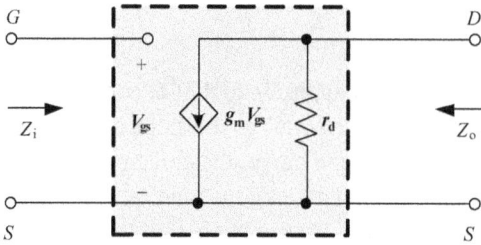

Fig. 4.68: FET AC equivalent circuit [11–13, 16].

V_{gs} to distinguish it from DC levels. In addition, the terminal S is common to both input and output loops, whereas terminals G and D are only in relationship through the controlled current source $g_m V_{gs}$.

Note that in the case when r_d is sufficiently large compared with other elements of the network, it can be replaced with an open circuit. So, the FET AC equivalent circuit is just a controlled current source $g_m V_{gs}$ with controlling voltage V_{gs} and parameter g_m.

4.7 JFET AC analysis

After the definition of the FET equivalent circuit, a number of fundamental JFET small-signal configurations are examined. The processes contain the determination of the important AC parameters of Z_i, Z_o and A_v for each configuration, which parallels the AC analysis of BJT amplifiers introduced in Chapter 3.

4.7.1 Fixed-bias configuration

The fixed-bias configuration of the p-channel JFET is shown in Fig. 4.69, which includes coupling capacitors C_1 and C_2, separating the DC biasing arrangement from the applied signal V_i and output V_o.

For small-signal AC analysis, parameters are added to show the network properties: the AC input signal V_i is applied to the terminal G of the JFET, whereas the AC output V_o is off the terminal D. The AC parameters of Z_i and Z_o are extracted from the input port and the output port, respectively [3, 11, 12].

Now modify the original network to its AC equivalent circuit: replace the DC blocking capacitors C_1 and C_2 by the short-circuit equivalents, because the reactance $X_c = 1/(2\pi f C)$ is sufficiently small compared to other element impedance of the network. Change the DC supply V_{DD} to ground and V_{GG} to short circuit. The AC equivalent circuit is shown in Fig. 4.70 [11, 12].

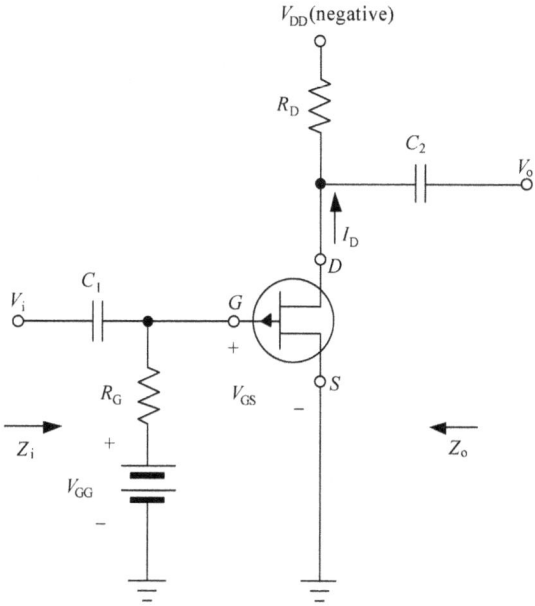

Fig. 4.69: AC analysis of fixed-bias configuration for the p-channel JFET.

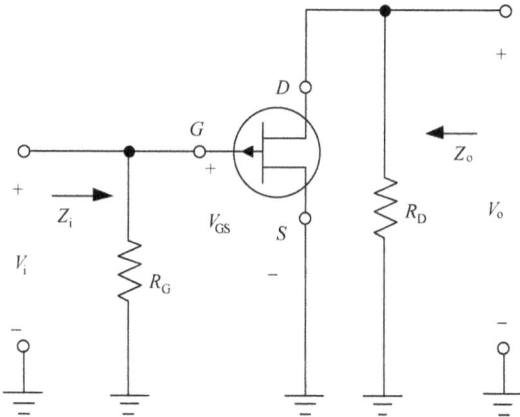

Fig. 4.70: AC equivalent circuit of fixed-bias configuration for the p-channel JFET.

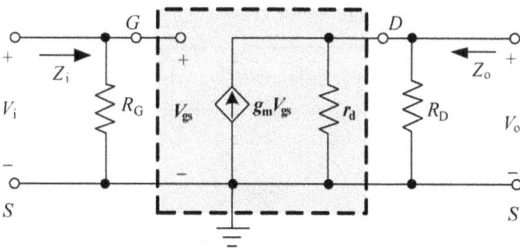

Fig. 4.71: AC equivalent model of fixed-bias configuration for the p-channel JFET.

After determination of g_m and r_d from the DC biasing arrangement, specification sheet or characteristics, the AC equivalent model can be substituted between the appropriate terminals as shown in Fig. 4.71. Note the direction of arrows in the current source, which indicates the realistic direction of the drain current of the p-channel JFET.

Figure 4.71 clearly shows that

$$Z_i = R_G$$

From the knowledge of circuit analysis, the output impedance of a network, Z_o, is defined as the equivalent impedance in the condition when all the voltage sources are set as short circuits and current sources as open-circuits. So in Fig. 4.71, when the controlled current source is replaced with the open-circuit equivalence, Z_o can be obtained:

$$Z_o = r_d \parallel R_D$$

In the case of $r_d \geq 10 R_D$, the following approximation is acceptable:

$$Z_o \approx R_D$$

In the output port, from Ohm's law we obtain

$$V_o = (g_m V_{gs})(r_d \parallel R_D)$$

and in the input port,

$$V_i = V_{gs}$$

So that

$$V_o = (g_m V_i)(r_d \parallel R_D)$$

Then, the voltage gain A_v is

$$A_v = \frac{V_o}{V_i} = g_m(r_d \parallel R_D) \tag{4.13}$$

In the case of $r_d \geq 10 R_D$, the following approximation is acceptable:

$$A_v = g_m R_D \tag{4.14}$$

For the common-source configuration, the negative polarity of A_v, resulting from negative g_m, reveals that a phase shift of 180° between input V_i and output V_o voltages. This is demonstrated in Fig. 4.72. Note that the amplitudes of V_o and V_i are opposite but with the same period of T, indicating that it is linear amplification [3, 10–12].

Example 4.10
The p-channel JFET fixed-bias configuration with the Q-point of $V_{GSQ} = 2$ V and $I_{DQ} = 3.6$ mA is shown in Fig. 4.73. Find the values of g_m, Z_i, Z_o and A_v.

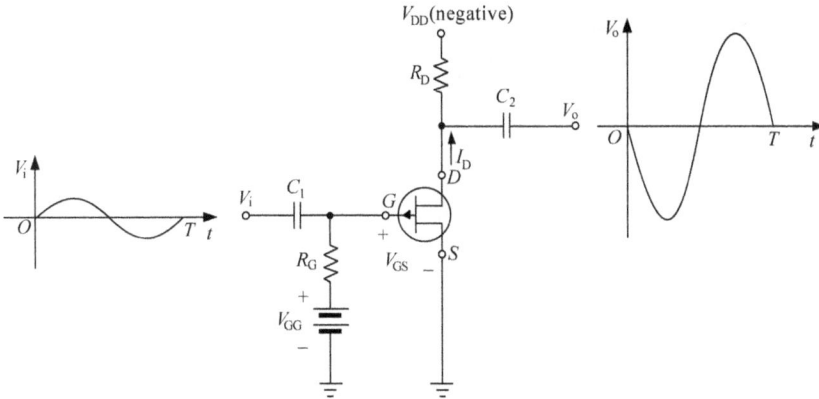

Fig. 4.72: 180° phase shift of fixed-bias configuration for the p-channel JFET.

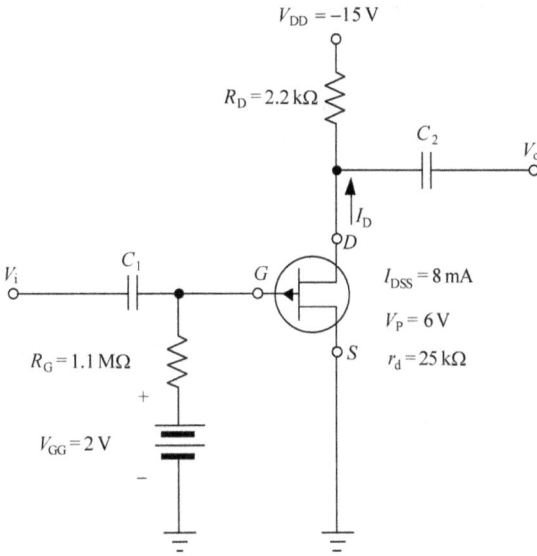

Fig. 4.73: Network of Example 4.10.

Solution

Use the fixed-bias configuration of the p-channel JFET AC equivalent model, as shown in Fig. 4.71.

For g_m: from Eq. (4.11), the transconductance g_m can be obtained analytically

$$g_m = \frac{2I_{DSS}}{V_P^2}(V_{GS} - V_P)$$

$$= \frac{2 \times 8\,mA}{(6\,V)^2}(2\,V - 6\,V)$$

$$= -1.78\,mS$$

For Z_i: from Fig. 4.71, it is clear that

$$Z_i = R_G$$
$$= 1.1\,\text{M}\Omega$$

For Z_o:

$$Z_o = r_d \parallel R_D$$
$$= 25\,\text{k}\Omega \parallel 2.2\,\text{k}\Omega$$
$$= 2.02\,\text{k}\Omega$$

For A_v: from Eq. (4.13), we obtain

$$A_v = -g_m(r_d \parallel R_D)$$
$$= (-1.78\,\text{mS}) \times (25\,\text{k}\Omega \parallel 2.2\,\text{k}\Omega)$$
$$= (-1.78\,\text{mS}) \times (2.02\,\text{k}\Omega)$$
$$= -3.6$$

From A_v, it can be concluded that the common-source configuration, with the p-channel JFET and the fixed-bias circuit, is a reversed linear amplification.

4.7.2 Self-bias configuration

The fixed-bias configuration discussed before requires two DC voltage sources, which is an obvious disadvantage. The self-bias configuration of Fig. 4.74 needs only one DC power supply to establish the desired Q-point.

For small-signal AC analysis, parameters are added to show the network properties: the AC input signal V_i is applied to the terminal G of the JFET, whereas the AC output V_o is off the terminal D. The AC parameters of Z_i and Z_o are extracted from the input port and the output port respectively.

Modify the original network to its AC equivalent circuit: replace the DC blocking capacitors C_1 and C_2 by the short-circuit equivalents. Change the DC supply V_{DD} to ground. The capacitor C_s across the source resistance R_s is assumed to be an open circuit for DC analysis, allowing R_s to set up the Q-point as discussed in Section 4.4.2. Under AC conditions, the capacitor C_s should be assumed as a short circuit to "remove" the effects of R_s. So, the AC equivalent circuit is shown in Fig. 4.75.

Then from the DC biasing arrangement, specification sheet or characteristics, g_m and r_d can be determined. The AC equivalent model can be substituted between the corresponding terminals as shown in Fig. 4.76. Note the direction of the arrow in the current source, coming out of the device, which indicates the realistic direction of the drain current of the p-channel JFET.

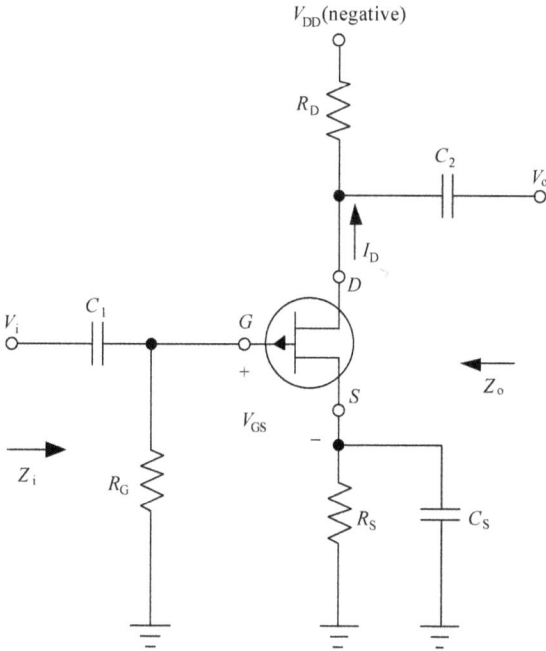

Fig. 4.74: AC analysis of self-bias configuration for the p-channel JFET.

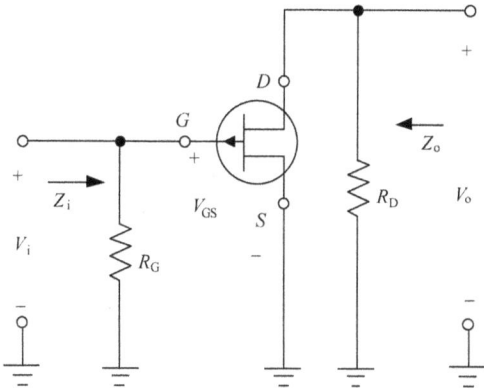

Fig. 4.75: AC equivalent circuit of self-bias configuration for the p-channel JFET.

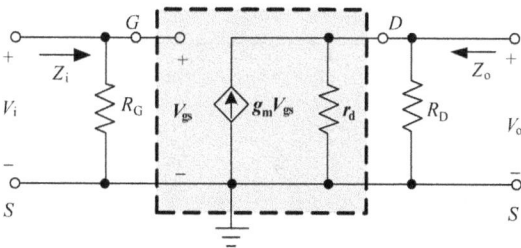

Fig. 4.76: AC equivalent model of self-bias configuration for the p-channel JFET.

Figure 4.76 clearly shows that

$$Z_i = R_G$$

and Z_o can be obtained

$$Z_o = r_d \parallel R_D$$

In the case of $r_d \geq 10R_D$, the following approximation is acceptable:

$$Z_o \approx R_D$$

In the output port, from Ohm's law we obtain

$$V_o = (g_m V_{gs})(r_d \parallel R_D)$$

In the input port,

$$V_i = V_{gs}$$

So that

$$V_o = (g_m V_i)(r_d \parallel R_D)$$

Then, the voltage gain A_v is

$$A_v = \frac{V_o}{V_i} = g_m(r_d \parallel R_D) \tag{4.15}$$

In the case of $r_d \geq 10R_D$, the following approximation is acceptable:

$$A_v = g_m R_D \tag{4.16}$$

Because g_m is negative for p-channel devices, A_v of the common-source configuration is negative, revealing that V_o and V_i are 180° out of phase.

Example 4.11
The p-channel JFET self-bias configuration with the Q-point of $V_{GSQ} = 3.1$ V and $I_{DQ} = 3.0$ mA is shown in Fig. 4.77. Find the values of g_m, Z_i, Z_o and A_v.

Solution
Use the self-bias configuration of the p-channel JFET AC equivalent model, as shown in Fig. 4.76.

For g_m: from Eq. (4.11), the transconductance g_m can be obtained analytically

$$g_m = \frac{2I_{DSS}}{V_P^2}(V_{GS} - V_P)$$

$$= \frac{2 \times 10\,\text{mA}}{(8\,\text{V})^2}(3.1\,\text{V} - 8\,\text{V})$$

$$= -1.53\,\text{mS}$$

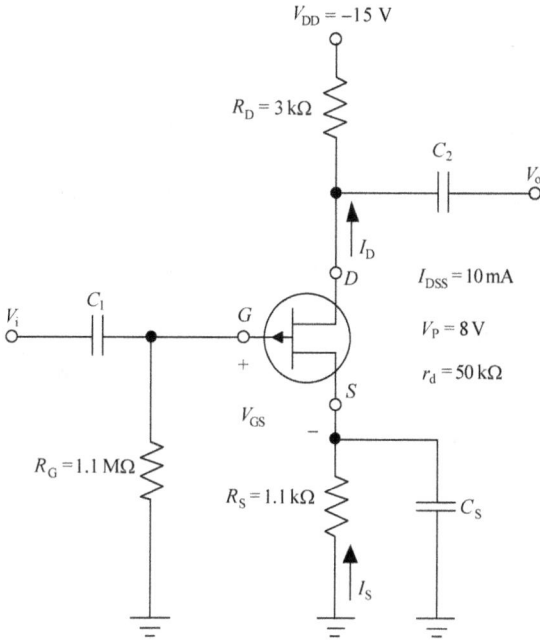

Fig. 4.77: Network of Example 4.11.

For Z_i: from Fig. 4.76, it is clear that

$$Z_i = R_G$$
$$= 1.1\,\text{M}\Omega$$

For Z_o:

$$Z_o = r_d \parallel R_D$$
$$= 50\,\text{k}\Omega \parallel 3\,\text{k}\Omega$$
$$= 2.83\,\text{k}\Omega$$

For A_v: from Eq. (4.15), we obtain

$$A_v = -g_m(r_d \parallel R_D)$$
$$= (-1.78\,\text{mS}) \times (50\,\text{k}\Omega \parallel 3\,\text{k}\Omega)$$
$$= (-1.78\,\text{mS}) \times (2.83\,\text{k}\Omega)$$
$$= -5.03$$

From A_v, it can be concluded that the common-source configuration, with the p-channel JFET and the self-bias circuit, is a reversed linear amplification. Note also that the typical gain of a JFET amplifier is less than that of BJTs with similar configurations. However, Z_i of a JFET amplifier is generally greater than the typical Z_i of a BJT, which will have a very positive effect on the overall gain of a system. This really accounts for the value of a JFET amplifier.

4.7.3 Voltage-divider configuration

The voltage-divider bias configuration, applied to BJT amplifiers in Section 3.4.5, can also be applied to p-channel JFET amplifiers, as illustrated in Fig. 4.78.

Replacing the DC supply V_{DD} with ground leads to that both the upper ends of R_{G1} and R_D are grounded. Since both R_{G1} and R_{G2} share a common ground, they can be connected in parallel; R_D can also be brought down to ground. Also, the capacitor C_s is assumed as the short circuit to "remove" the effects of R_s. The AC equivalent circuit is shown in Fig. 4.79.

After determination of g_m and r_d from the DC biasing arrangement, specification sheet or characteristics, the AC equivalent model can be substituted between the corresponding terminals as shown in Fig. 4.80. Note the direction of the arrow of the controlled current source, which indicates the realistic direction of the drain current of the p-channel JFET. Figure 4.80 clearly shows that

$$Z_i = R_{G1} \parallel R_{G2}$$

and Z_o can be obtained.

$$Z_o = r_d \parallel R_D$$

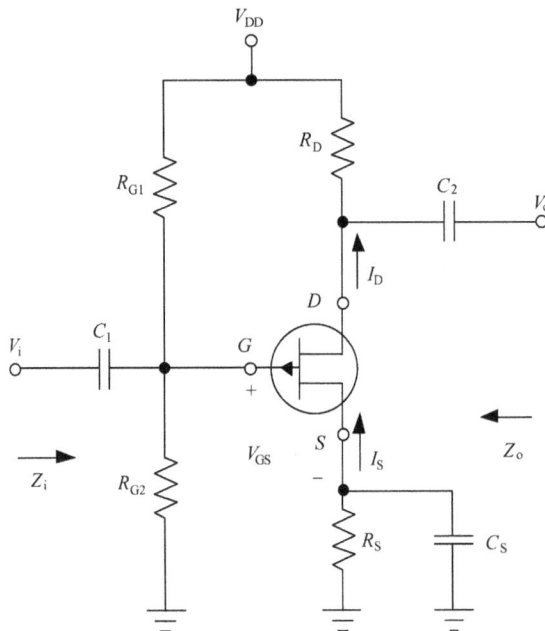

Fig. 4.78: AC analysis of voltage-divider configuration for the p-channel JFET.

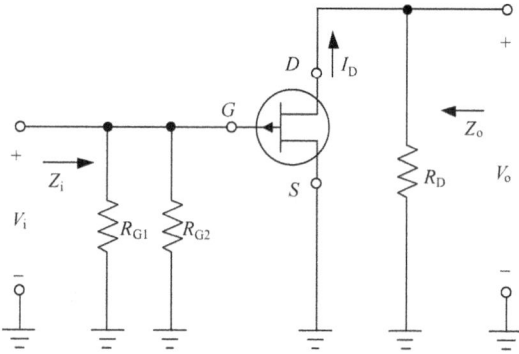

Fig. 4.79: AC equivalent circuit of voltage-divider configuration for the p-channel JFET.

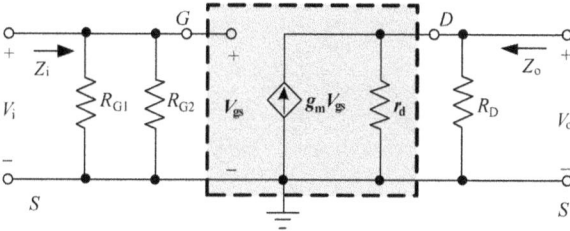

Fig. 4.80: AC equivalent model of voltage-divider configuration for the p-channel JFET.

In the output port, from Ohm's law we obtain

$$V_o = (g_m V_{gs})(r_d \parallel R_D)$$

and in the input port,

$$V_i = V_{gs}$$

So that

$$V_o = (g_m V_i)(r_d \parallel R_D)$$

Then the voltage gain A_V is

$$A_V = \frac{V_o}{V_i} = g_m(r_d \parallel R_D) \tag{4.17}$$

Because g_m is negative for p-channel devices, A_V of the common-source configuration is negative, revealing that V_o and V_i are 180° out of phase.

Note that the equations of the fixed-bias, self-bias and voltage-divider configurations for Z_o and A_V are obtained in the same way. The only difference is the equation of the voltage divider configuration for Z_i, which is the parallel combination of R_1 and R_2.

4.7.4 Source-follower configuration

The JFET equivalent of the BJT emitter-follower configuration is the source follower, as shown in Fig. 4.81.

Substituting the JFET equivalent circuit results in the configuration of Fig. 4.82.

The controlled source $g_m V_{gs}$ and the internal output impedance r_d of the JFET are tied to terminal D, actually AC ground, at one end and R_S on the other, with V_o across R_S. Since $g_m V_{gs}$, r_d and R_S are connected to the same terminal and ground, the ground can be moved down as usual. Then $g_m V_{gs}$ and r_d can all be placed in parallel as shown in Fig. 4.83. The current source $g_m V_{gs}$ has its direction pointing down, but V_{gs} is still defined between the G and S terminals.

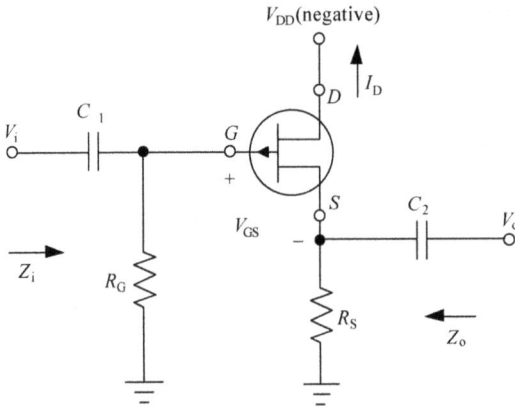

Fig. 4.81: AC analysis of source-follower configuration for p-channel JFET.

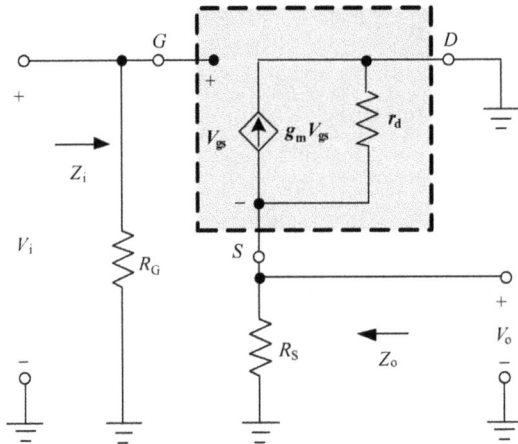

Fig. 4.82: AC equivalent model of the source-follower configuration for the p-channel JFET.

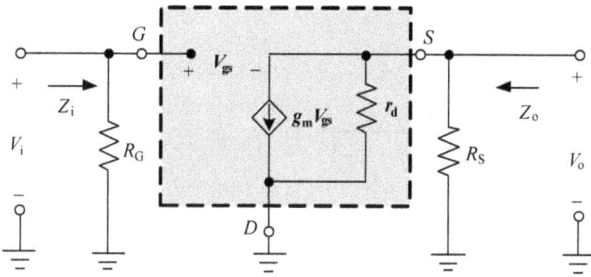

Fig. 4.83: Rearranged AC equivalent model of the source-follower configuration.

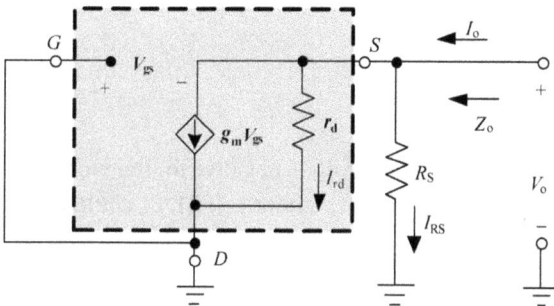

Fig. 4.84: Calculating Z_o of the source-follower configuration.

For Z_i: from Fig. 4.83 we obtain

$$Z_i = R_G$$

For Z_o:

Setting $V_i = 0\,V$ leads to the G terminal being connected directly to the ground, as shown in Fig. 4.84. The fact is clear that V_{gs} and V_o are across the same parallel network. So,

$$V_o = -V_{gs}$$

Applying Kirchhoff's current law at node S,

$$\begin{aligned}
I_o &= I_{r_d} + I_{R_S} + g_m V_{gs} \\
&= \frac{V_o}{r_d} + \frac{V_o}{R_S} + g_m V_{gs} \\
&= \frac{V_o}{r_d} + \frac{V_o}{R_S} + g_m(-V_o) \\
&= V_o\left(\frac{1}{r_d} + \frac{1}{R_S} - g_m\right)
\end{aligned}$$

Then,

$$Z_0 = \frac{V_0}{I_0}$$

$$= \frac{V_0}{V_0 \left(\frac{1}{r_d} + \frac{1}{R_S} - g_m \right)}$$

$$= \frac{1}{\frac{1}{r_d} + \frac{1}{R_S} - g_m}$$

$$= \frac{1}{\frac{1}{r_d} + \frac{1}{R_S} + \frac{1}{1/(-g_m)}}$$

It looks like the Z_0 is the equivalent resistance of three resistors in parallel connection. Therefore, in a more convenient form,

$$Z_0 = r_d \parallel R_S \parallel \frac{1}{-g_m}$$

Also note that it has been concluded before that g_m is negative for the p-channel device and positive for the n-channel one. So, for the n-channel JFET, Z_0 could be

$$Z_0 = r_d \parallel R_S \parallel (1/g_m)$$

For A_v:

In the output port, from Ohm's law we obtain

$$V_0 = (-g_m V_{gs})(r_d \parallel R_S)$$

From Fig. 4.83 it is obvious that the difference between the input port and output port is V_{gs}, so,

$$V_i - V_0 = V_{gs}$$

So that,

$$V_0 = (-g_m)(V_i - V_0)(r_d \parallel R_S)$$
$$= (-g_m)V_i(r_d \parallel R_S) - (-g_m)V_0(r_d \parallel R_S)$$

and

$$V_0 + (-g_m)V_0(r_d \parallel R_S) = (-g_m)V_i(r_d \parallel R_S)$$

That is,

$$V_0(1 + (-g_m)(r_d \parallel R_S)) = (-g_m)V_i(r_d \parallel R_S)$$

So that,

$$A_v = \frac{V_0}{V_i} = \frac{(-g_m)(r_d \parallel R_S)}{1 + (-g_m)(r_d \parallel R_S)} \tag{4.18}$$

It can be seen that the numerator and denominator of Eq. (4.18) are almost same, so the result will be near to 1 and smaller than 1. Moreover, A_v of the source-follower configuration is positive, revealing that V_0 and V_i are in phase.

Example 4.12

The p-channel JFET source-follower configuration with the Q-point of $V_{GSQ} = 2.75$ V and $I_{DQ} = 4.62$ mA is shown in Fig. 4.85. Find the values of g_m, Z_i, Z_o and A_v.

Solution

As shown in Fig. 4.83, use the source-follower configuration of the p-channel JFET AC equivalent model.

For g_m: from Eq. (4.11), the transconductance g_m can be obtained analytically

$$g_m = \frac{2I_{DSS}}{V_P^2}(V_{GS} - V_P)$$

$$= \frac{2 \times 15 \text{ mA}}{(5 \text{ V})^2}(2.75 \text{ V} - 5 \text{ V})$$

$$= -2.7 \text{ mS}$$

For Z_i: from Fig. 4.83, it is clear that

$$Z_i = R_G = 1.1 \text{ M}\Omega$$

For Z_o:

$$Z_o = r_d \parallel R_S \parallel (1/g_m)$$

$$= \frac{1}{\frac{1}{40 \text{ k}\Omega} + \frac{1}{2 \text{ k}\Omega} + \frac{1}{1/(2.7 \text{ mS})}}$$

$$= 310 \ \Omega$$

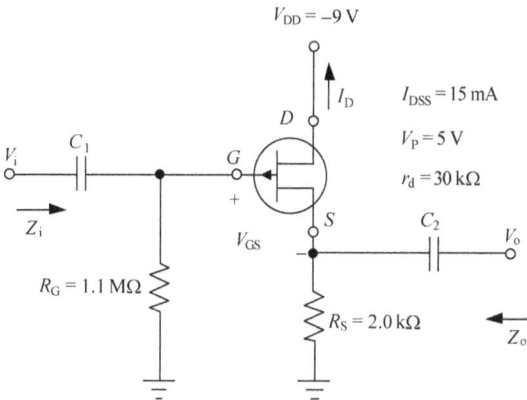

Fig. 4.85: Network of Example 4.12.

For A_V: from Eq. (4.15), we obtain

$$A_v = \frac{(-g_m)(r_d \parallel R_S)}{1 + (-g_m)(r_d \parallel R_S)}$$
$$= \frac{(2.7\,\text{mS})(40\,\text{k}\Omega \parallel 2\,\text{k}\Omega)}{1 + (2.7\,\text{mS})(40\,\text{k}\Omega \parallel 2\,\text{k}\Omega)}$$
$$= \frac{(2.7\,\text{mS})(1.90\,\text{k}\Omega)}{1 + (2.7\,\text{mS})(1.90\,\text{k}\Omega)}$$
$$= \frac{5.13}{1 + 5.13}$$
$$= 0.84$$

4.7.5 Common-gate configuration

The common-gate configuration parallels the common-base configuration of BJTs discussed in Section 3.4.8. The network is shown in Fig. 4.86.

Substituting the JFET equivalent circuit results in the configuration of Fig. 4.87. The controlled source $g_m V_{gs}$ is connected from S to D with the internal output impedance r_d in parallel; R_S is across the input loop and R_D is across the output loop (since V_{DD} can be regarded as AC ground). The terminal G is common ground. Note that V_{gs} is opposite to V_i and both of them are across R_S.

An auxiliary variable Z_i' is introduced to facilitate the determination of Z_i, as shown in Fig. 4.87. The network to find Z_i' is illustrated in Fig. 4.88.

Applying Kirchhoff's voltage law in the network leads to

$$V' = V_{rd} + V_{RD}$$

Then,

$$V_{rd} = V' - V_{RD}$$

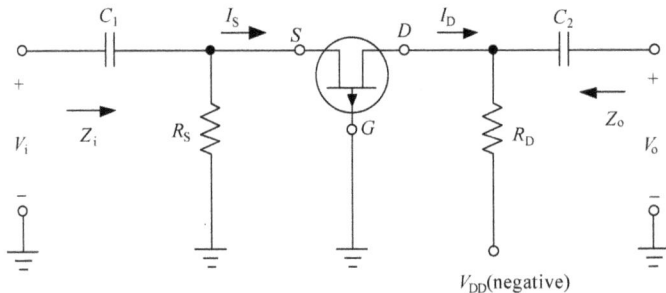

Fig. 4.86: AC analysis of common-gate configuration for the p-channel JFET.

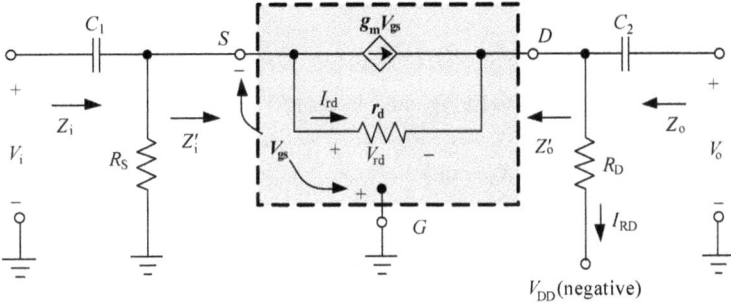

Fig. 4.87: AC equivalent model of common-gate configuration for the p-channel JFET.

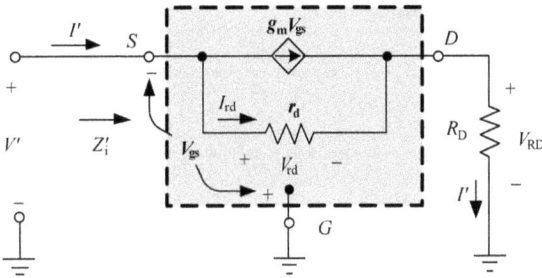

Fig. 4.88: Network to find auxiliary variable Z_i'.

Also, applying Kirchhoff's current law at node of left end of resistor r_d leads to

$$I' = g_m V_{gs} + I_{rd}$$

$$= g_m V_{gs} + \frac{V_{rd}}{r_d}$$

$$= g_m(-V') + \frac{V' - V_{RD}}{r_d}$$

$$= V'\left(-g_m + \frac{1}{r_d}\right) - \frac{V_{RD}}{r_d}$$

$$= V'\left(-g_m + \frac{1}{r_d}\right) - \frac{I' R_D}{r_d}$$

Then,

$$I'\left(1 + \frac{R_D}{r_d}\right) = V'\left(-g_m + \frac{1}{r_d}\right)$$

So,

$$Z_i' = \frac{V'}{I'}$$

$$= \frac{1 + \frac{R_D}{r_d}}{-g_m + \frac{1}{r_d}}$$

$$= \frac{r_d + R_D}{1 - g_m r_d}$$

Finally,

$$Z_i = R_S \parallel Z_i'$$

To determine Z_o, setting V_i to zero in Fig. 4.87, leads to V_{gs} to zero and removal of R_S; r_d and R_D remain. Then,

$$Z_o = r_d \parallel R_D$$

Also as illustrated in Fig. 4.87, we obtain

$$V_i = -V_{gs}$$

and

$$V_o = I_{RD} R_D$$

Moreover, the voltage across r_d is

$$V_{rd} = V_i - V_o$$

So,

$$I_{rd} = \frac{V_i - V_o}{r_d}$$

Applying Kirchhoff's current law at the node of the right end of the resistor r_d leads to

$$I_{RD} = g_m V_{gs} + I_{rd} = g_m(-V_i) + \frac{V_i - V_o}{r_d}$$

So,

$$V_o = I_{RD} R_D$$
$$= \left(g_m(-V_i) + \frac{V_i - V_o}{r_d} \right) R_D$$
$$= g_m(-V_i)R_D + \frac{V_i}{r_d} R_D + \frac{-V_o}{r_d} R_D$$

Then,

$$V_o \left(1 + \frac{R_D}{r_d} \right) = V_i \left(-g_m R_D + \frac{R_D}{r_d} \right)$$

So,

$$A_v = \frac{V_o}{V_i}$$
$$= \frac{-g_m R_D + \frac{R_D}{r_d}}{1 + \frac{R_D}{r_d}}$$
$$= \frac{\frac{1}{r_d} - g_m}{\frac{1}{r_d} + \frac{1}{R_D}}$$

Note that g_m will still be negative for the p-channel device and positive for the n-channel one. Moreover, A_v of the common-gate configuration is positive, revealing that V_o and V_i are in phase.

Example 4.13

The p-channel JFET common-gate configuration with the Q-point of $V_{GSQ} = 2.3$ V and $I_{DQ} = 2.1$ mA is shown in Fig. 4.89. Find the values of g_m, Z_i, Z_o and A_v.

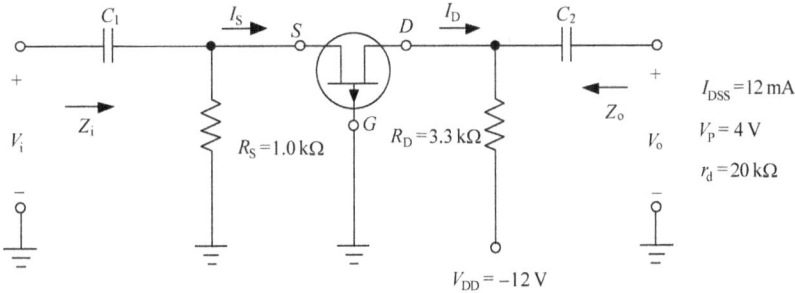

Fig. 4.89: Network of Example 4.13.

Solution

As shown in Fig. 4.87, use the common-gate configuration of the p-channel JFET AC equivalent model.

For g_m: from Eq. (4.11), the transconductance g_m can be obtained analytically

$$g_m = \frac{2I_{DSS}}{V_P^2}(V_{GS} - V_P)$$

$$= \frac{2 \times 12 \text{ mA}}{(4 \text{ V})^2}(2.3 \text{ V} - 4 \text{ V})$$

$$= -2.55 \text{ mS}$$

For Z_i: from Fig. 4.83, it is clear that

$$Z_i = R_S \parallel Z_i'$$

$$= R_S \left\| \frac{r_d + R_D}{1 - g_m r_d} \right.$$

$$= 1.0 \text{ k}\Omega \left\| \frac{20 \text{ k}\Omega + 3.3 \text{ k}\Omega}{1 - (-2.55 \text{ mS}) \times 20 \text{ k}\Omega} \right.$$

$$= 1.0 \text{ k}\Omega \parallel 0.45 \text{ k}\Omega$$

$$= 0.31 \text{ k}\Omega$$

For Z_o:

$$Z_o = r_d \parallel R_D$$

$$= 20 \text{ k}\Omega \parallel 3.3 \text{ k}\Omega$$

$$= 2.83 \text{ k}\Omega$$

For A_V: from Eq. (4.15), we obtain

$$
A_V = \frac{\frac{1}{r_d} - g_m}{\frac{1}{r_d} + \frac{1}{R_D}}
$$

$$
= \frac{\frac{1}{20\,k\Omega} - (-2.55\,mS)}{\frac{1}{20\,k\Omega} + \frac{1}{3.3\,k\Omega}}
$$

$$
= \frac{0.05\,mS + 2.55\,mS}{0.05\,mS + 0.30\,mS}
$$

$$
= 7.43
$$

4.8 MOSFET AC analysis

4.8.1 Depletion-type MOSFETs

As discussed in Section 4.2, most parts of the transfer characteristics of depletion-type MOSFETs are similar to those of JFETs, except that V_{GS} of depletion-type MOSFETs can be extended to be negative for the p-channel and positive for the n-channel, and I_D greater than I_{DSS}.

Because the transfer characteristics of depletion-type MOSFETs conform to Shockley's equation, Eq. (4.11), for transconductance g_m is still applicable, and the AC equivalent model of depletion-type MOSFETs is similar to that of JFETs. The range of internal resistance r_d is quite similar to that of JFETs [1, 3].

Example 4.14

The p-channel depletion-type MOSFET voltage-divider configuration with the Q-point of $V_{GSQ} = -0.15\,V$ and $I_{DQ} = 10.5\,mA$ (in Fig. 4.54) is shown in Fig. 4.90. Find the values of g_m, Z_i, Z_o and A_V.

Solution

As shown in Fig. 4.91, use the voltage-divider configuration of the p-channel depletion-type MOSFET AC equivalent model. Note that R_S has been "removed" by C_S in the AC analysis.

For g_m: from Eq. (4.11), the transconductance g_m can be obtained analytically,

$$
g_m = \frac{2I_{DSS}}{V_P^2}(V_{GS} - V_P)
$$

$$
= \frac{2 \times 10\,mA}{(6\,V)^2}(-0.15\,V - 6\,V)
$$

$$
= -3.42\,mS
$$

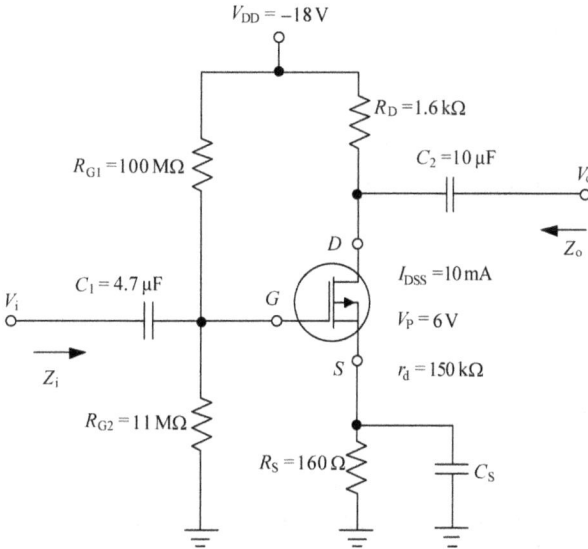

Fig. 4.90: Network of Example 4.14.

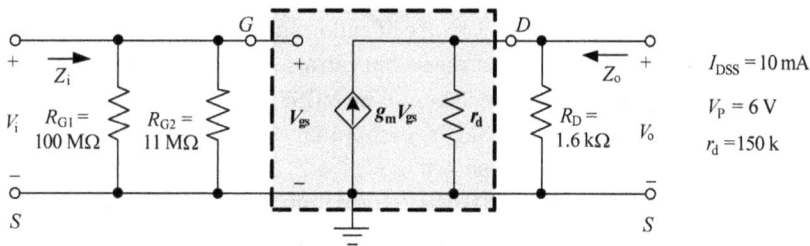

Fig. 4.91: AC equivalent model of voltage-divider configuration for the *p*-channel depletion-type MOSFET.

For Z_i: from Fig. 4.91, the same as discussed in Section 4.7.3 for the JFET, we obtain

$$Z_i = R_{G1} \parallel R_{G1}$$
$$= 100\,\text{M}\Omega \parallel 11\,\text{M}\Omega$$
$$= 9.91\,\text{M}\Omega$$

For Z_o:

$$Z_o = r_d \parallel R_D$$
$$= 150\,\text{k}\Omega \parallel 1.6\,\text{k}\Omega$$
$$= 1.58\,\text{k}\Omega$$

From Eq. (4.17), the voltage gain A_V is

$$A_V = \frac{V_o}{V_i} = g_m(r_d \parallel R_D)$$

$$= (-3.42 \, \text{mS}) \times (150 \, \text{k}\Omega \parallel 1.6 \, \text{k}\Omega)$$

$$= (-3.42 \, \text{mS}) \times (1.58 \, \text{k}\Omega)$$

$$= -5.41$$

A_V of the common-source configuration for depletion-type MOSFET is negative, revealing that V_o and V_i are 180° out of phase.

4.8.2 Enhancement-type MOSFET

The AC small-signal equivalent circuit of the enhancement-type MOSFET, discussed already in Section 4.6.4, is shown again in Fig. 4.92. It can be seen that there is an open-circuit between G and S; and a current source exists between D and S, the magnitude of which dependents on the gate-to-source voltage Vgs, and the direction of which indicates the device type, n-channel or p-channel. There is output impedance r_d from D to S, which is usually provided on specification sheets. The device transconductance g_m is provided on specification sheets but can also be found analytically. For AC analysis in most respects, the enhancement-type MOSFET is the same as that employed for JFETs or depletion-type MOSFETs, except that the transfer characteristics only exist beyond the threshold V_T, as shown in Figs. 4.23 and 4.25 [1, 3, 11, 16].

In previous analysis of FETs, Eq. (4.11) was derived from Shockley's equation to obtain the transconductance g_m analytically. For enhancement-type MOSFETs, the relationship between output current I_D and controlling voltage V_{GS} is defined by Eq. (4.2) as follows:

$$I_D = k \, (V_{GS} - V_T)^2$$

Since g_m is still defined by Eq. (4.10) as

$$g_m = \frac{\Delta I_D}{\Delta V_{GS}}$$

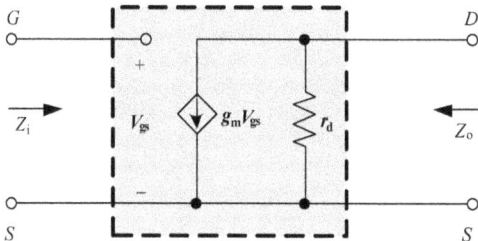

Fig. 4.92: Enhancement-type MOSFET AC equivalent model.

Then,

$$g_m = \frac{dI_D}{dV_{GS}}\Big|_{Q\text{-point}}$$

$$= \frac{d}{dV_{GS}}\left[k(V_{GS} - V_T)^2\right]$$

$$= 2k(V_{GS} - V_T)\frac{d}{dV_{GS}}[V_{GS} - V_T]$$

$$= 2k(V_{GS} - V_T)$$

That is,

$$g_m = 2k(V_{GSQ} - V_T) \tag{4.19}$$

where k is the parameter described by Eq. (4.3).

1. Voltage-divider configuration

The voltage-divider bias configuration, applied to BJT amplifiers in Section 3.4.5 and p-channel JFET amplifiers in Section 4.7.3 can also be applied to p-channel enhance-ment-type MOSFET amplifiers, as illustrated in Fig. 4.93.

Replacing the DC supply V_{DD} with ground leads to that both the upper ends of R_{G1} and R_D are grounded. Since both R_{G1} and R_{G2} share a common ground, they can be connected in parallel. R_D can also be brought down to ground. Also, the capacitor C_s is assumed as the short circuit to "remove" the effects of R_s. The AC equivalent circuit is shown in Fig. 4.94. Note that the drain current I_D comes of the device as for p-channel enhancement-type MOSFETs.

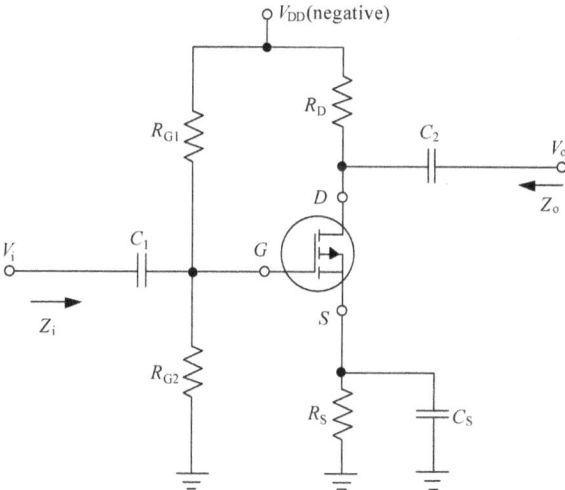

Fig. 4.93: AC analysis of voltage-divider configuration for p-channel enhancement-type MOSFET amplifiers.

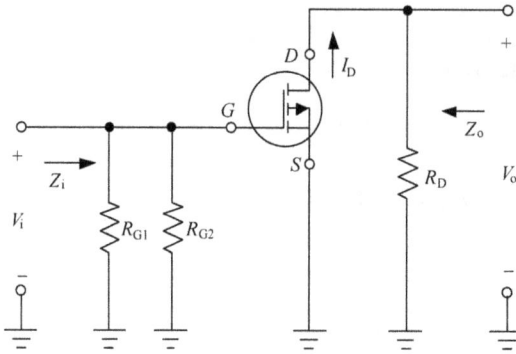

Fig. 4.94: AC equivalent circuit of voltage-divider configuration for the p-channel enhancement-type MOSFET.

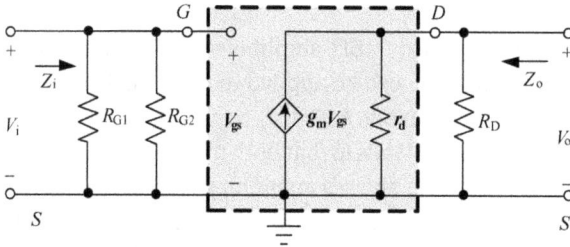

Fig. 4.95: AC equivalent model of voltage-divider configuration for the p-channel enhancement-type MOSFET.

After determination of g_m and r_d from the DC biasing arrangement, specification sheet or characteristics, the AC equivalent model can be substituted between the corresponding terminals as shown in Fig. 4.95. Note the direction of arrow of the controlled current source, which indicates the realistic direction of the drain current of the p-channel JFET.

Figure 4.95 clearly shows that

$$Z_i = R_{G1} \parallel R_{G2}$$

and Z_o can be obtained

$$Z_o = r_d \parallel R_D$$

In the output port, from Ohm's law we obtain

$$V_o = (g_m V_{gs})(r_d \parallel R_D)$$

In the input port,

$$V_i = V_{gs}$$

So that

$$V_o = (g_m V_i)(r_d \parallel R_D)$$

Then, the voltage gain A_V is

$$A_V = \frac{V_o}{V_i} = g_m(r_d \parallel R_D) \tag{4.20}$$

Because g_m is negative for the p-channel enhancement-type MOSFET, A_V of the common-source amplifier is negative, revealing that V_o and V_i are 180° out of phase.

Moreover, for the n-channel enhancement-type MOSFET, the direction of the arrow of the controlled current source in the AC model points down, which is opposite to that in Fig. 4.95.

Further,

$$A_V = -g_m(r_d \parallel R_D)$$

2. Drain-feedback configuration

The drain-feedback configuration, of which DC analysis was carried out in Section 4.5.2, is the focus of AC analysis now, as shown in Fig. 4.96.

In the drain-feedback configuration, the feedback resistor R_G provides a sufficiently-high impedance between V_i and V_o, avoiding that the output and input are at the same level. Substituting the AC equivalent model for the device leads to the network of Fig. 4.97.

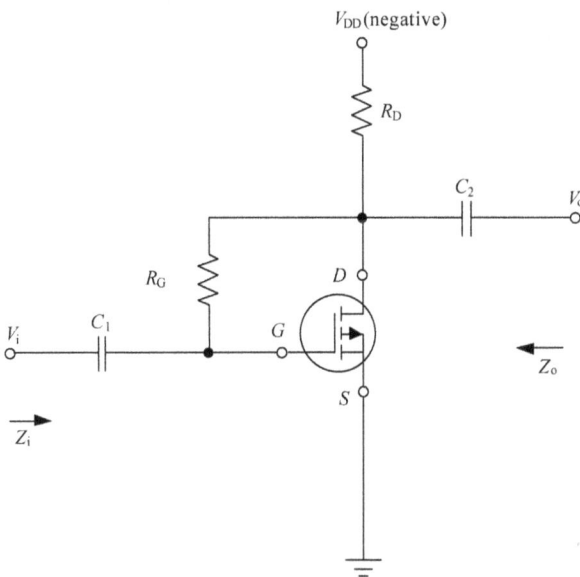

Fig. 4.96: AC analysis of drain-feedback configuration for the p-channel enhancement-type MOSFET.

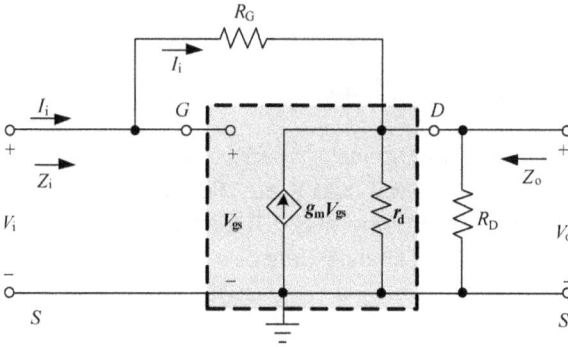

Fig. 4.97: AC equivalent model of drain-feedback configuration for the p-channel enhancement-type MOSFET.

Applying Kirchhoff's current law to the right end of R_G, we obtain

$$I_i + g_m V_{gs} = \frac{V_o}{r_d \parallel R_D}$$

and

$$V_{gs} = V_i$$

So that,

$$I_i + g_m V_i = \frac{V_o}{r_d \parallel R_D}$$

That is,

$$V_o = (r_d \parallel R_D)(I_i + g_m V_i)$$

On the other hand,

$$I_i = \frac{V_i - V_o}{R_G}$$

and

$$I_i = \frac{V_i - V_o}{R_G}$$
$$= \frac{V_i - (r_d \parallel R_D)(I_i + g_m V_i)}{R_G}$$

Arranging the equation to separate V_i and I_i, we obtain

$$Z_i = \frac{V_i}{I_i}$$
$$= \frac{R_G + (r_d \parallel R_D)}{1 - g_m(r_d \parallel R_D)}$$

To find Z_o, set V_i to zero, leading to $V_{gs} = 0\,\text{V}$ and $g_m V_{gs} = 0\,\text{V}$. Therefore, in the network, only R_G, r_d and R_D remain in parallel connection, as illustrated in Fig. 4.98. Thus,

$$Z_o = R_G \parallel r_d \parallel R_D$$

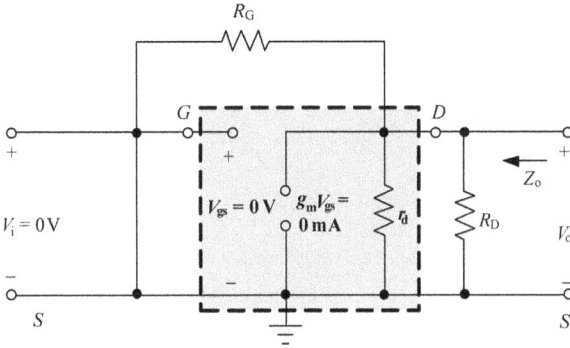

Fig. 4.98: Network to find Z_o.

In Fig. 4.97, by applying Kirchhoff's current law to the right end of R_G, it has already been obtained that

$$I_i + g_m V_{gs} = \frac{V_o}{r_d \parallel R_D}$$

with

$$V_{gs} = V_i$$

and

$$I_i = \frac{V_i - V_o}{R_G}$$

So that,

$$\frac{V_i - V_o}{R_G} + g_m V_i = \frac{V_o}{r_d \parallel R_D}$$

Arranging the equation to separate V_i and V_o, we obtain

$$A_v = \frac{V_o}{V_i} = \frac{\frac{1}{R_G} + g_m}{\frac{1}{R_G} + \frac{1}{r_d \parallel R_D}}$$

Also, note that

$$\frac{1}{R_G} + \frac{1}{r_d \parallel R_D} = \frac{1}{R_G \parallel r_d \parallel R_D}$$

Finally,

$$A_v = \left(\frac{1}{R_G} + g_m\right)(R_G \parallel r_d \parallel R_D)$$

Because g_m is negative for the p-channel enhancement-type MOSFET, and

$$|g_m| \gg \frac{1}{R_G}$$

the A_v of the drain-feedback amplifier is negative, revealing that V_o and V_i are 180° out of phase.

Example 4.15

The p-channel enhancement-type MOSFET amplifier with drain-feedback bias configuration with $k = 0.25 \times 10^{-3}$ A/V^2, $V_{GSQ} = -6.4$ V and $I_{DQ} = 2.6$ mA is shown in Fig. 4.99. Find the values of g_m, Z_i, Z_o and A_v.

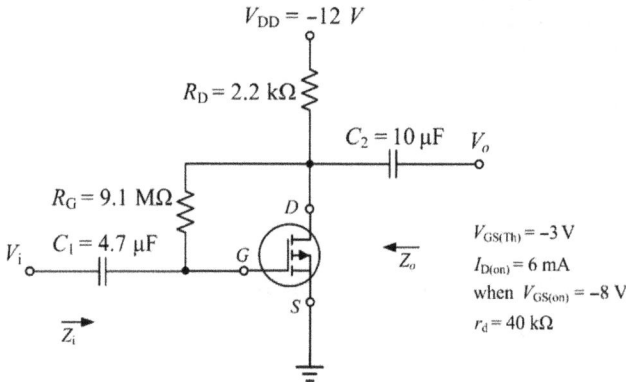

$V_{DD} = -12$ V

$R_D = 2.2$ kΩ

$C_2 = 10$ μF V_o

$R_G = 9.1$ MΩ

V_i $C_1 = 4.7$ μF G D Z_o

S Z_i

$V_{GS(Th)} = -3$ V
$I_{D(on)} = 6$ mA
when $V_{GS(on)} = -8$ V
$r_d = 40$ kΩ

Fig. 4.99: Network of Example 4.15.

Solution

For g_m: from Eq. (4.19), the transconductance g_m can be obtained analytically

$$g_m = 2k(V_{GSQ} - V_T)$$
$$= 2 \times 0.25 \times 10^{-3} \text{A/V}^2 \times (-6.4 \text{ V} - (-3 \text{ V}))$$
$$= -1.7 \text{ mS}$$

For Z_i:

$$Z_i = \frac{R_G + (r_d \parallel R_D)}{1 - g_m(r_d \parallel R_D)}$$
$$= \frac{9.1 \text{ MΩ} + (40 \text{ kΩ} \parallel 2.2 \text{ kΩ})}{1 - (-1.7 \text{ mS})(40 \text{ kΩ} \parallel 2.2 \text{ kΩ})}$$
$$= \frac{9.1 \text{ MΩ} + 2.09 \text{ kΩ}}{1 - (-1.7 \text{ mS}) \times 2.09 \text{ kΩ}}$$
$$= \frac{9.1 \text{ MΩ} + 2.09 \text{ kΩ}}{1 + 3.55}$$
$$= 2.0 \text{ MΩ}$$

For Z_o:

$$Z_o = R_G \parallel r_d \parallel R_D$$
$$= 9.1 \text{ MΩ} \parallel 40 \text{ kΩ} \parallel 2.2 \text{ kΩ}$$
$$= 2.08 \text{ kΩ}$$

For A_V:

$$A_v = \left(\frac{1}{R_G} + g_m\right)(R_G \parallel r_d \parallel R_D)$$

$$= \left[\frac{1}{9.1\,\text{M}\Omega} + (-1.7\,\text{mS})\right] \times (9.1\,\text{M}\Omega \parallel 40\,\text{k}\Omega \parallel 2.2\,\text{k}\Omega)$$

$$\approx (-1.7\,\text{mS}) \times (2.08\,\text{k}\Omega)$$

$$= -3.54$$

The value of A_V shows that the drain-feedback amplifier is also a reversed amplifier.

4.9 Chapter summary

4.9.1 JFET and MOSFET Summary

Symbols
JFET graphic symbols:

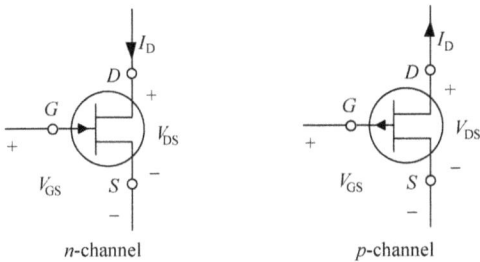

n-channel   *p*-channel

Fig. 4.100: JFET graphic symbols.

Depletion-type MOSFET graphic symbols:

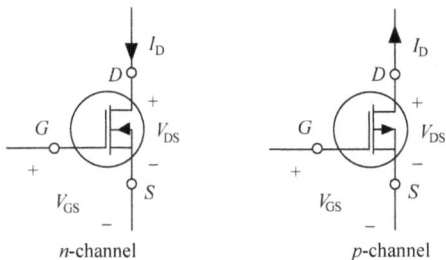

n-channel   *p*-channel

Fig. 4.101: Depletion-type MOSFET graphic symbols.

Enhancement-type MOSFET graphic symbols:

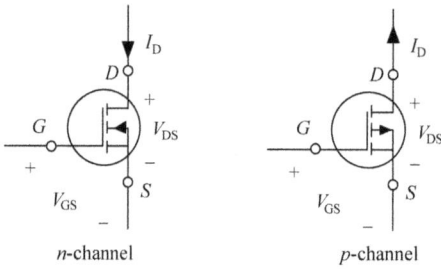

Fig. 4.102: Enhancement-type MOSFET graphic symbols.

Comparison between JFETs and BJTs

The JFET transistor is a voltage-controlled device, whereas the BJT is a current-controlled device. See Chapter 4.

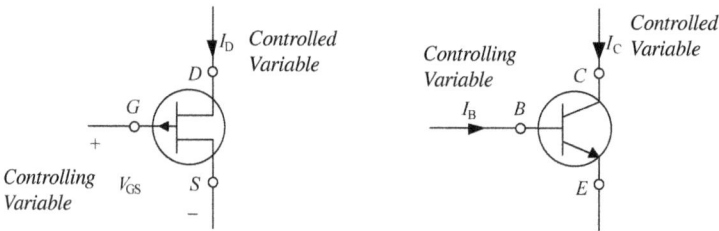

Fig. 4.103: Comparison between JFET and BJT.

Relationships

JFET		BJT	
$I_D = I_{DSS} \left(1 - \frac{V_{GS}}{V_P}\right)^2$	Eq. (4.1)	$I_C = \beta I_B$	Eq. (3.7)
$I_D = I_S$	Eq. (4.4)	$I_E \approx I_C$	Eq. (3.2)
$I_D \approx 0\,A$	Eq. (4.5)	$V_{BE} = 0.7\,V$	Eq. (3.3)

Concepts and conclusions

(1) The maximum current for JFET is labeled I_{DSS} and occurs when $V_{GS} = 0\,V$. See Section 4.1.2.

(2) The minimum current for a JFET occurs at pinch-off condition defined by $V_{GS} = V_P$. See Section 4.1.2.

(3) The relationship between the drain current I_D and the gate-to-source voltage V_{GS} of a JFET is a nonlinear one defined by Shockley's equation. As the current level approaches I_{DSS}, the sensitivity of I_D to changes in V_{GS} increases significantly. See Section 4.1.3.

(4) The transfer characteristics ($I_D \sim V_{GS}$) are characteristics of the device itself and are not sensitive to the network in which the JFET is employed. See Section 4.1.3.

(5) The four points can be used to sketch the transfer curve to a satisfactory level of accuracy. See Tab. 4.1.

(6) The arrow in the symbol of the n-channel JFET will always point to the center of the symbol, whereas that of a p-channel device will always point out of the center of the symbol. See Fig. 4.10.

(7) MOSFETs are available in one of two types: depletion and enhancement. See Sections 4.2 and 4.3.

(8) The depletion-type MOSFET has the same transfer characteristics as a JFET for drain currents up to the I_{DSS} level. At this point the characteristics of a depletion-type MOSFET continue to levels above I_{DSS} whereas those of the JFET will end. See Section 4.2.2.

(9) The arrow in the symbol of the n-channel MOSFET will always point to the center of the symbol, whereas those of a p-channel device will always point out of the center of the symbol. See Fig. 4.18.

(10) The transfer characteristics of an enhancement-type MOSFET are defined by a nonlinear equation controlled by V_{GS}, V_T and k. The resulting plot of $I_D \sim V_{GS}$ rises exponentially with increasing values of V_{GS}. See Section 4.3.2.

Equations

For JFETs and depletion-type MOSFETs

$$I_D = I_{DSS} \left(1 - \frac{V_{GS}}{V_P} \right)^2 \qquad \text{Eq. (4.1)}$$

For enhancement-type MOSFETs

$$I_D = k\,(V_{GS} - V_T)^2 \qquad \text{Eq. (4.2)}$$

$$k = \frac{I_{D(on)}}{(V_{GS(on)} - V_T)^2} \qquad \text{Eq. (4.3)}$$

4.9.2 DC biasing summary

Concepts and conclusions
(1) The JFET fixed-bias configuration has a fixed DC voltage applied from G to S to establish the Q-point. See Section 4.4.1.
(2) The nonlinear relationship between the V_{GS} and I_D of a JFET requires that a graphical or mathematical solution be used to determine the Q-point. See Section 4.4.1.
(3) The JFET self-bias configuration has a load line that will always pass through the origin. See Section 4.4.2.
(4) For the JFET common-gate configuration, the load line $I_D \sim V_{GS}$ function defines a straight line through the origin. See Section 4.4.3.
(5) For the JFET voltage-divider biasing configuration, one can always assume that I_G is zero to permit an isolation of the input loop from the output loop. See Section 4.4.4.
(6) The resulting V_G voltage will always be positive for an n-channel JFET and negative for a p-channel JFET. See Section 4.4.4.
(7) Increasing values of R_S results in lower I_D. See Section 4.4.4.
(8) The method of DC analysis applied to depletion-type MOSFETs is the same as that applied to JFETs, with the only difference being a possible Q-point with an I_D above I_{DSS}. See Section 4.5.1.
(9) The method of DC analysis applied to enhancement-type MOSFETs contains the same steps as for depletion-type MOSFETs, except that the transfer characteristics of an enhancement-type MOSFET are defined by a nonlinear equation controlled by V_{GS}, V_T and k. For values of V_{GS} less than the threshold value V_T, the drain current I_D is zero. See Section 4.5.2.

Equations
For the JFET fixed-bias configuration:

$$V_{GS} = V_{GG}$$

For the JFET self-bias configuration:

$$V_{GS} = I_D R_S$$

For the JFET common-base configuration:

$$V_{GS} = I_D R_S$$

For the JFET voltage-divider configuration:

$$V_G = \frac{R_{G2}}{R_{G1} + R_{G2}} V_{DD}$$
$$V_{GS} = V_G + R_S I_D$$

For the depletion-type MOSFET voltage-divider configuration:

$$V_G = \frac{R_{G2}}{R_{G1} + R_{G2}} V_{DD}$$

For the depletion-type MOSFET self-bias configuration:

$$V_{GS} = I_D R_S$$

For the depletion-type MOSFET fixed-bias configuration:

$$V_{GS} = 0$$

For the enhancement-type MOSFET voltage-divider configuration:

$$V_G = \frac{R_{G2}}{R_{G1} + R_{G2}} V_{DD}$$

For the enhancement-type MOSFET drain-feedback configuration:

$$V_{GS} = V_{DD} + I_D R_D$$

Transfer characteristics of JFETs:

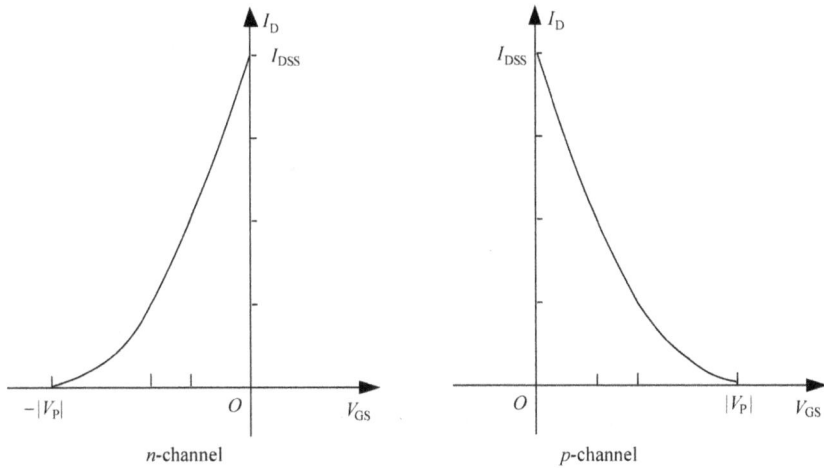

Fig. 4.104: Transfer characteristics of JFETs.

Transfer characteristics of depletion-type MOSFETs:

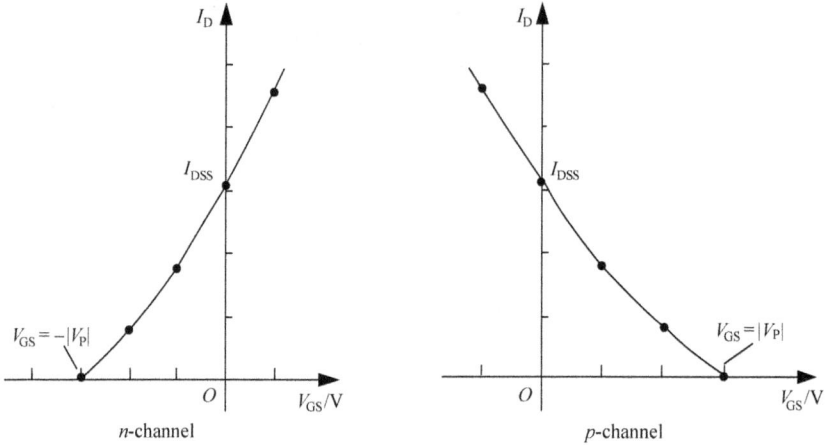

Fig. 4.105: Transfer characteristics of depletion-type MOSFETs.

Transfer characteristics of enhancement-type MOSFET:

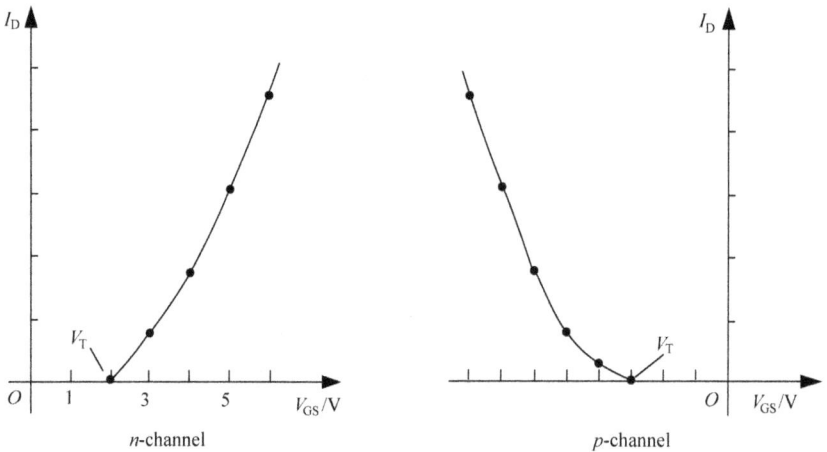

Fig. 4.106: Transfer characteristics of enhancement-type MOSFETs.

4.9.3 AC analysis summary

Concepts and conclusions

(1) The transconductance g_m is determined by the ratio of the change in I_D associated with a particular change in V_{GS}. The steeper the slope of the $I_D \sim V_{GS}$ curve, the greater the level of g_m. See Section 4.6.1.

(2) The input impedance Z_i of commonly used FETs is so large that is assumed as an open circuit. See Section 4.6.2.

(3) The output impedance Z_o of FETs is similar in magnitude to that of conventional BJTs. It is defined on the drain characteristics as the slope of the characteristic curve at the Q-point. See Section 4.6.3.

(4) The more horizontal the characteristic curves on the drain characteristics, the greater the output impedance r_d. See Section 4.6.4.

(5) A model for the FET in the AC domain can be set up as in Section 4.6.4.

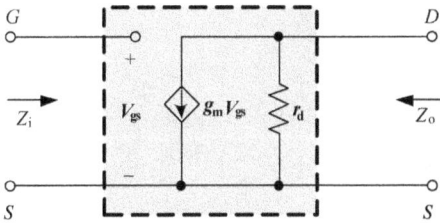

Fig. 4.107: AC model for the FET.

(6) The voltage gain for the fixed-bias and self-bias JFET configurations are same. See Sections 4.7.1 and 4.7.2.

(7) The AC analysis of JFETs and depletion-type MOSFETs is the same. See Sections 4.7 and 4.8.

(8) The AC equivalent circuit for an enhancement-type MOSFET is the same as that employed for JFET and depletion-type MOSFET. The only difference is the equation for g_m. See Section 4.8.2.

(9) The magnitude of the gain of FET networks is typically between 2 and 20. See the examples in Sections 4.7 and 4.8.

(10) For the source-follower and common-gate configurations, there is no phase shift between input V_i and output V_o. Most others are reversed amplifiers. See the examples in Sections 4.7 and 4.8.

(11) The output impedance Z_o for most FET configurations is determined mainly by R_D. For the source-follower configuration it is determined by R_S and g_m. See the examples in Sections 4.7 and 4.8.

(12) The input impedance Z_i for most FET configurations is very high. However, it is very low for the common-gate configuration. See the examples in Sections 4.7 and 4.8.

Equations

Definition of transconductance g_m:

$$g_m = \frac{\Delta I_D}{\Delta V_{GS}}$$

Eq. (4.10)

Transconductance g_m for JFETs and depletion-type MOSFETs:

$$g_m = \frac{2I_{DSS}}{V_P^2}(V_{GS} - V_P)$$

Eq. (4.11)

Transconductance g_m for enhancement-type MOSFETs:

$$g_m = 2k(V_{GSQ} - V_T)$$

Eq. (4.19)

For JFET amplifiers with fixed-bias configuration:

$$A_v = \frac{V_o}{V_i} = g_m(r_d \parallel R_D)$$

Eq. (4.13)

For JFET amplifiers with self-bias configuration:

$$A_v = \frac{V_o}{V_i} = g_m(r_d \parallel R_D)$$

Eq. (4.15)

For JFET amplifiers with voltage-divider configuration:

$$A_v = \frac{V_o}{V_i} = g_m(r_d \parallel R_D)$$

Eq. (4.17)

For JFET amplifiers with source-follower configuration:

$$A_v = \frac{V_o}{V_i} = \frac{(-g_m)(r_d \parallel R_S)}{1 + (-g_m)(r_d \parallel R_S)} \approx 1$$

Eq. (4.18)

For JFET amplifiers with common-gate configuration:

$$A_v = \frac{\frac{1}{r_d} - g_m}{\frac{1}{r_d} + \frac{1}{R_D}}$$

For depletion-type MOSFET amplifiers with voltage-divider configuration:

$$A_v = \frac{V_o}{V_i} = g_m(r_d \parallel R_D)$$

For enhancement-type MOSFET amplifiers with voltage-divider configuration:

$$A_v = \frac{V_o}{V_i} = g_m(r_d \parallel R_D)$$

Eq. (4.20)

4.10 Questions

Q4.1: Find the main difference between JFETs and BJTs from the viewpoint of controlling and controlled parameters.

Q4.2: List the main features of FETs compared with BJTs.

Q4.3: Plot the basic construction of n- and p-channel JFETs. Find the same and different parts between them.

Q4.4: Explain how biasing voltage controls the channel width and the magnitude of current for JFETs.

Q4.5: How do engineers tackle the difficulty of using the nonlinear relationship of Eq. (4.1)?

Q4.6: How are the transfer curves created from drain characteristics?

Q4.7: Plot the drain and transfer characteristics of n- and p-channel JFETs. Find the difference between them.

Q4.8: Explain the principle of the graphical approach to obtain the Q-point, as shown in Fig. 4.30.

Q4.9: Plot the basic construction of n- and p-channels for D-MOSFET and E-MOSFET. Find the same and different parts between them.

Q4.10: Explain how biasing voltage controls the channel and the magnitude of current for MOSFETs.

Q4.11: Draw the drain and transfer characteristics of n- and p-channel D-MOSFETs and E-MOSFETs. Find the difference between them.

Q4.12: There are 15 examples in Chapter 4, most of which involve p-channel devices. Try to solve the same questions with n-channel devices, thus find the common and different properties for those devices.

Q4.13: The JFET circuit is shown in Fig. 4.108; the parameters are known: $g_m = -1\,\text{mS}$, $R_1 = 330\,\text{k}\Omega$, $R_2 = 110\,\text{k}\Omega$, $R_3 = 1.1\,\text{M}\Omega$, $R_4 = 11\,\text{k}\Omega$, $R_5 = 2\,\text{k}\Omega$, $R_6 = 11\,\text{k}\Omega$. Estimate A_v, Z_i and Z_o.

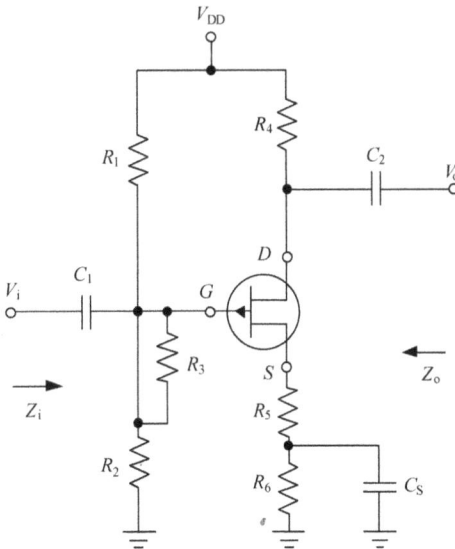

Fig. 4.108: Network of Q4.11.

5 Operational amplifiers and their applications

5.1 Overview

An operational amplifier, usually abbreviated to op-amp or opamp, is a DC-coupled high-gain electronic voltage amplifier with a differential input and, normally, a single-ended output. Under this configuration, an op-amp generates an output voltage, typically hundreds of thousands of times larger than the voltage difference between its two input terminals. Operational amplifiers had their original application in analog computers for many linear and nonlinear mathematical operations [13].

The versatility of the op-amp is the reason for its popularity in analog circuits as a building block. With the help of negative feedback, the characteristics of an op-amp circuit, such as gain, bandwidth, input impedance, output impedance and so on, are manipulated by external components and have almost nothing to do with temperature change or manufacturing variations [12].

Packaged as components or used as elements of more complex integrated circuits, op-amps are one of the most widely used electronic devices today, especially in the field of analog circuits, for a wide range of applications in industrial, scientific and consumer devices. Many standard IC op-amps cost only a few cents; while some integrated or hybrid operational amplifiers with special functions may cost over US$ 100 [12].

Moreover, the op-amp is one type of differential amplifier. Other types of differential amplifier include the fully differential amplifier, the instrumentation amplifier, the isolation amplifier and the negative-feedback amplifier. However, some of these are beyond the coverage of the textbook.

Briefly, the following is the historical timeline of the op-amps [12, 21]. In the 1940s, an op-amp was first invented as a "summing amplifier" by Swartzel of Bell Labs in 1941. Throughout World War II, this invention demonstrated its value, which would have been impossible for other devices. In 1947, an operational amplifier was designed by Loebe Julie. It had two major innovations: in its input stage, a long-tailed triode pair was used with matched loads to reduce drift in the output; its two inputs, one inverting, the other noninverting, made a whole range of new functionality possible. In 1949, a chopper-stabilized op-amp was invented by Goldberg, which involved a normal op-amp with an additional AC amplifier. This configuration significantly improved the op-amp gain, while greatly reducing the DC offset and output drift.

In the 1950s, Philbrick Research Incorporated released a vacuum tube op-amp in 1953, which was the first commercially available model of an op-amp. This op-amp started the widespread applications of op-amps in industry.

In the 1960s, solid-state, discrete op-amps were manufactured in 1961. These op-amps were actually miniature circuits on semiconductor substrates. In 1963, the first monolithic IC op-amp was invented by Widlar from Fairchild Semiconductor.

https://doi.org/10.1515/9783110593860-005

In the 1970s, a single-sided supply for op-amps was introduced in 1972. This resulted in the elimination of the need for a separate negative power supply, which was convenient for many applications. A quad package was also introduced into op-amp in 1972, that is, four separate op-amps in one package, which undoubtedly became an industry standard.

In the 1980s, MOSFETs were largely used in the manufacture of op-amps, leading to great improvements of the performance of op-amps.

In recent decades, with further development of the semiconductor industry, characteristics, performance and specifications of op-amps have been persistently improved. Although their basic application has changed due to the appearance of digital computers, these amplifiers are still widely used in various electronic devices. The term of "operational" means some mathematical operations on signal. However, op-amp itself performs no other operations without external elements, but signal amplification, which is its dominant function. Modern op-amps perform this function perfectly.

Nowadays, manufacturers produce a large number, several hundred most probably, of various op-amps; therefore, even simple enumeration of their parameters and characteristics, such as input or output resistance, is a certain problem. It is difficult to orient designers in this abundance of types and parameters without the necessary structured knowledge about them. Therefore, the correct and reasonable choice of an op-amp for some device should include the cost, reliability and quality of the device under development.

5.2 Fundamentals of op-amp

An operational amplifier is a differential amplifier with a very high gain, featuring in high input impedance and low output impedance. There are many functional circuits built by op-amps, which are widely used to deal with analog signals. Moreover, an op-amp may contain multiple differential amplifier stages to achieve very high voltage gain [1, 9].

For applications of op-amps, the common opinion is that it is not necessary to know their internal circuits; on the contrary, it is sufficient to be aware of the input and output characteristics. This way of thinking is not rare. For example, computer manufacturers scarcely know the internal structure of the Pentium microprocessor, but this does not prevent them from designing satisfactory computers. Op-amps can be considered as circuit elements like resistors or capacitors, with only a somewhat more complex internal structure.

Nevertheless, the system modeling, involving the knowledge of the structure, structural relations and the principles of construction of various op-amps, allows engineers to design more competently. However, this is beyond the coverage of the textbook.

So, in the following sections, the basic knowledge of op-amps will be examined as a black box and the focus will be on the set up of the network to meet design requirements.

A basic op-amp is shown in Fig. 5.1 with two input terminals and one output terminal. The input terminal indicated by a plus (+) sign results in an output with the same polarity as that of input, while the one with minus (−) opposite polarity of input signal [1, 9, 11, 16].

Note that in Fig. 5.1, the terminals of power supply and other auxiliary pins are omitted for simplicity.

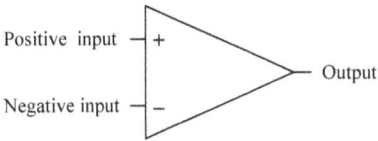

Fig. 5.1: Basic op-amp.

1. Single-ended input mode
When one input terminal is fed with an input signal while the other is connected to the ground, the op-amp is in single-ended input mode. As there are two input terminals, there are two types of single-ended input modes. When the positive input terminal is fed with a signal, and a negative input terminal is connected to the ground, as illustrated in Fig. 5.2, the output has the same polarity as the applied input signal [9, 11].

On the other hand, when the negative input terminal is fed with a signal and a positive input terminal is connected to ground, as illustrated in Fig. 5.3, the output has the opposite polarity to the applied input signal.

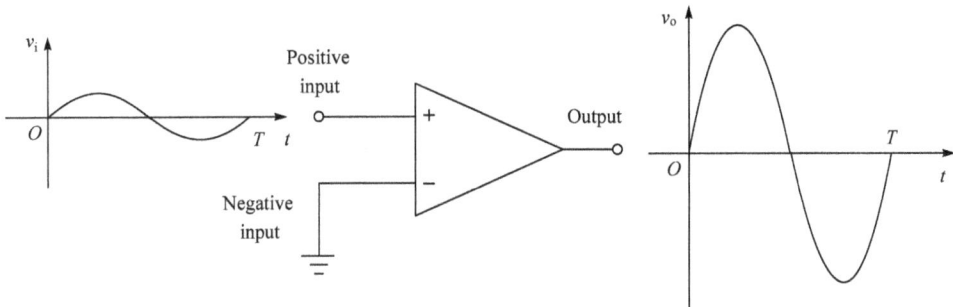

Fig. 5.2: Single-ended input operation mode with in-phase output signal.

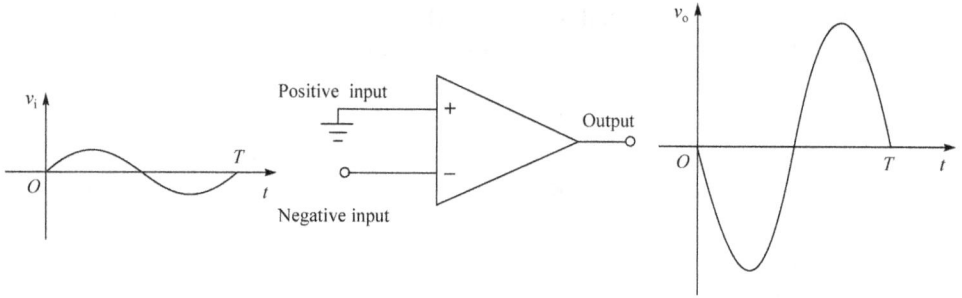

Fig. 5.3: Single-ended input operation mode with reversed output signal.

2. Double-ended (differential) input mode

When two input terminals of the op-amp are used for input signals, the op-amp is in double-ended input mode. Also, two different ways of usage of input terminals leads to two types of double-ended input modes, as shown in Figs. 5.4 and 5.5 [1, 9, 11, 16].

In Fig. 5.4, two separate signals, with respect to the ground, are applied to the input terminals, resulting in the amplification of difference signal of $v_{i1} - v_{i2}$.

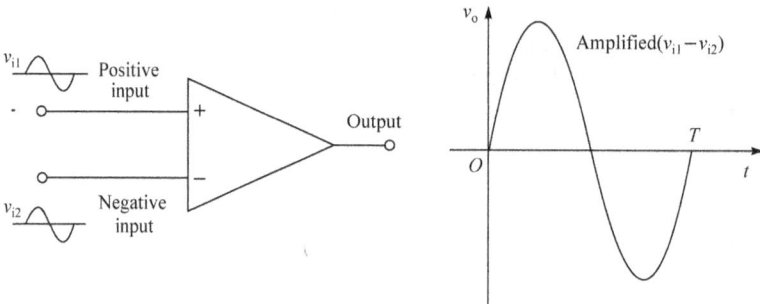

Fig. 5.4: Double-ended input operation mode with two input signals.

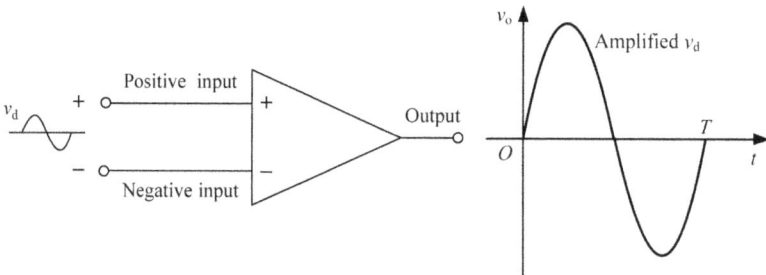

Fig. 5.5: Double-ended input operation mode with one input signal.

Figure 5.5 shows an input signal of v_d applied between the two input terminals, leading to amplified output in phase with v_d between the positive and negative inputs.

After the discussion of the input modes of op-amps, the output mode is introduced.

3. Double-ended output mode

The op-amp can also have two output terminals with opposite outputs, as shown in Fig. 5.6. Note that there is one more plus (+) and minus (–) sign inside the op-amp triangular symbol near the output terminals, representing the polarity of the output signals. Also the small circle at the beginning of the output terminal means that it is the output with opposite polarity with respect to the positive input terminal.

Fig. 5.7 shows a single-ended input with double-ended outputs [1, 9, 11, 16]. It can be seen that the signal applied to the positive input terminal leads to two amplified outputs. The one from the output terminal with a minus sign has opposite polarity to

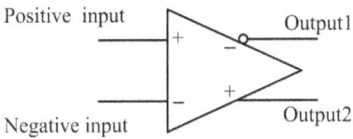

Fig. 5.6: Op-amp with double-ended outputs.

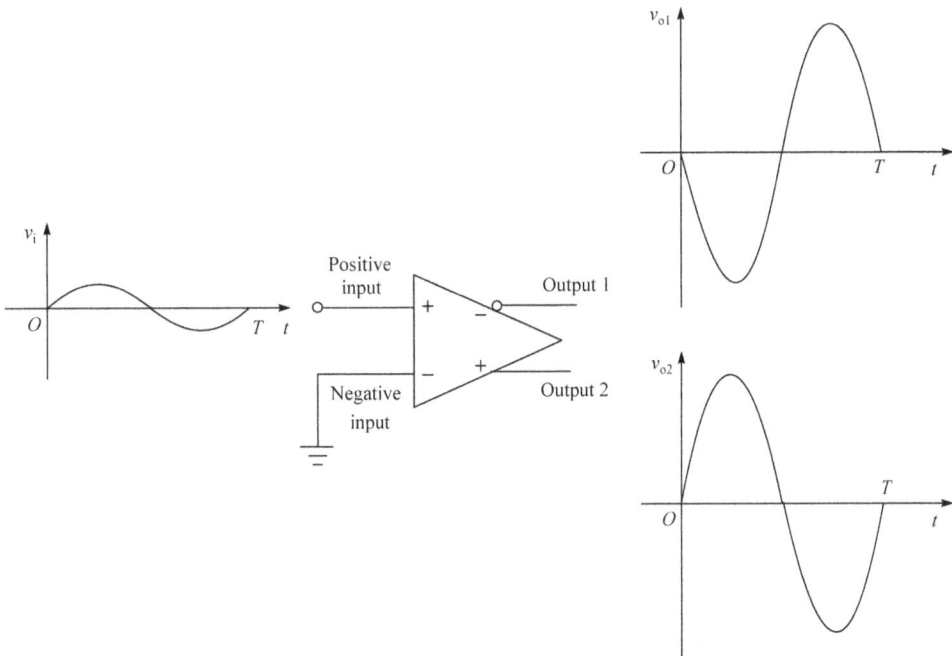

Fig. 5.7: Single-ended input with double-ended outputs.

the input signal, and the other from the output terminal with a plus sign has same polarity to the input signal. If the input signal has been changed to negative input terminal, all the polarity mentioned above should be changed respectively. Note that in this mode, all the input and output signals are with respect to ground.

Figure 5.8 shows the single-ended input operation mode with an input signal at the negative input terminal. At the output port, the output signal, extracted between the two output terminals, is not with respect to ground, and is referred to as a floating signal. The magnitude of this different output signal is amplified $v_{o1} - v_{o2}$. Note that v_{o1} and v_{o2} are opposite to each other, so the result of $v_{o1} - v_{o2}$ will be a signal of doubled v_{o1}.

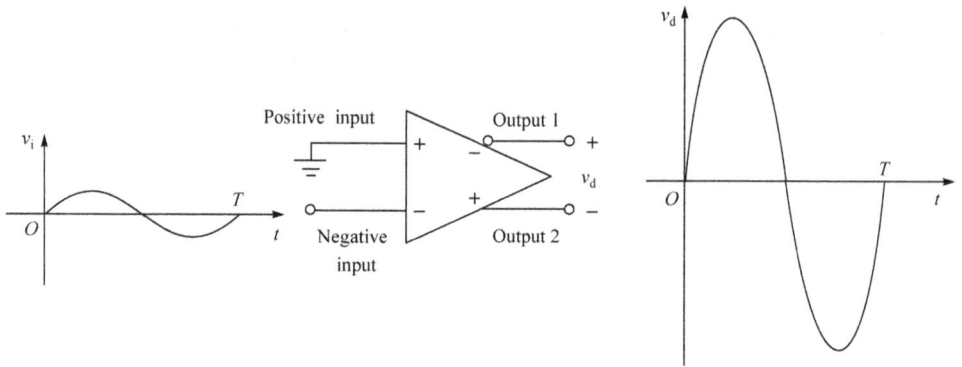

Fig. 5.8: Single-ended input with double-ended outputs between two output terminals.

Figure 5.9 shows a differential input with a differential output operation [1, 9, 11, 16]. The input is applied between the two input terminals and the output extracted from between the two output terminals. This is a fully differential operation. Both the input and the output signals are floating signals because they are not with respect to ground.

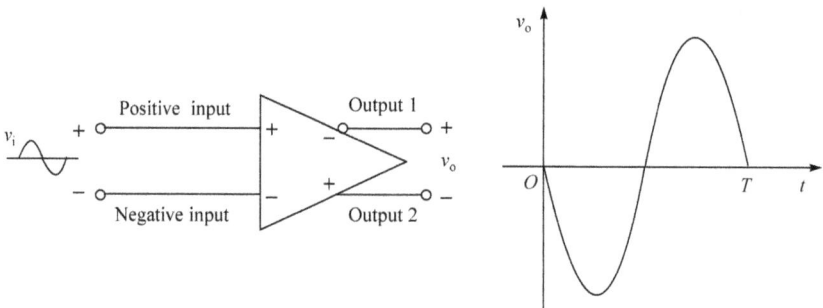

Fig. 5.9: Differential input with differential output operation.

4. Common-mode operation

When the same input signals are applied to both inputs, the op-amp is in the common-mode operation mode, as shown in Fig. 5.10. Ideally, the two inputs are equally amplified, leading to output signals with difference in polarity only at the output terminal. So, the final output signal is their cancellation, resulting in zero volt output. However, a small output signal will remain in practical conditions [1, 9, 11, 16].

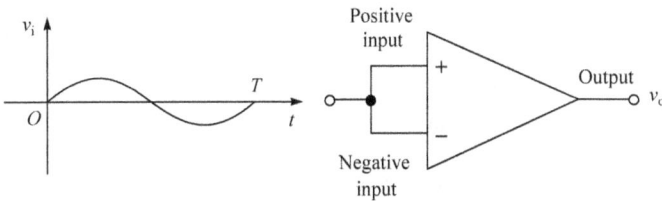

Fig. 5.10: Common-mode operation.

5. Common-mode rejection

An important feature when op-amp is in differential input mode, as in Fig. 5.4, is that the signals that are common at both inputs are only slightly amplified, while those that are opposite to the two inputs are highly amplified. In other words, the overall effect is to amplify the difference of the two input signals while cancelling the common signal at the two inputs [11, 16].

This has some practical significance when noise or any other types of interference is generally common to both inputs. The differential mode will attenuate these unwanted components at the input terminals while amplifying the different components at the input terminals. This operating feature is referred to as common-mode rejection.

5.3 Common-mode rejection ratio

In this section, some import mathematical equations relating to common-mode rejection are introduced to show this important parameter of the op-amp.

One of the most important features of an op-amp is the ability to largely amplify signal components that are opposite at the two input terminals, whereas only slightly amplifying signal components that are common to both input terminals. In other words, the amplification of an op-amp includes two parts: differential and common. Their values for an op-amp are considerably different, which can be described by the common-mode rejection ratio (CMRR). The following steps show the derivation of CMRR [11, 16].

When the independent inputs are fed to the op-amp, the difference signal component is the difference between the two inputs, which is called differential input voltage.

$$v_d = v_{i1} - v_{i2} \tag{5.1}$$

When both input signals are the same, a common signal component from the two inputs can be defined as the average of the two signals, which is called the common input voltage.

$$v_c = \frac{1}{2}(v_{i1} + v_{i2}) \tag{5.2}$$

Generally at the output port, any signal applied to an op-amp has both common and differential components. So, the consequent output can be expressed as

$$v_o = A_d v_d + A_c v_c \tag{5.3}$$

where A_d is differential gain of the op-amp and A_c is common-mode gain of the op-amp.

If input signals fed to an op-amp are have exactly opposite polarities, that is,

$$v_{i1} = -v_{i2} = v_s$$

the resulting difference voltage is

$$v_d = v_{i1} - v_{i2}$$
$$= v_s - (-v_s)$$
$$= 2v_s$$

and the resulting common voltage is

$$v_c = \frac{1}{2}(v_{i1} + v_{i2})$$
$$= \frac{1}{2}(v_s + (-v_s))$$
$$= 0$$

So that the overall output voltage is

$$v_o = A_d v_d + A_c v_c$$
$$= A_d(2v_s) + A_c \cdot 0$$
$$= 2A_d v_s$$

This indicates that when the inputs have ideally opposite polarities, that is, no common component, the output is the doubled product of differential gain and the applied input signal.

If input signals fed to an op-amp are have exactly the same polarities, that is,

$$v_{i1} = v_{i2} = v_s$$

the resulting difference voltage is

$$v_d = v_{i1} - v_{i2}$$
$$= v_s - v_s$$
$$= 0$$

and the resulting common voltage is

$$v_c = \frac{1}{2}(v_{i1} + v_{i2})$$
$$= \frac{1}{2}(v_s + v_s)$$
$$= v_s$$

So that the overall output voltage is

$$v_o = A_d v_d + A_c v_c$$
$$= A_d \cdot 0 + A_c(v_s)$$
$$= A_c v_s$$

This indicates that when the inputs have ideally common polarities, that is, no different component, the output is the product of the common-mode gain and the applied input signal. This means that only the common-mode operation exists.

Then the above-mentioned equations can be used to calculate A_d and A_c.

For A_d:

Set

$$v_{i1} = -v_{i2} = v_s = 0.5 \text{ V}$$

So that

$$v_d = v_{i1} - v_{i2}$$
$$= v_s - (-v_s)$$
$$= 2v_s$$
$$= 1 \text{ V}$$

and

$$v_c = \frac{1}{2}(v_{i1} + v_{i2})$$
$$= \frac{1}{2}(0.5 \text{ V} + (-0.5 \text{ V}))$$
$$= 0 \text{ V}$$

Then the output voltage is

$$v_o = A_d v_d + A_c v_c$$
$$= A_d \cdot (1 \text{ V}) + A_c \cdot (0 \text{ V})$$
$$= A_d$$

So that it is clear that by setting the input voltages $v_{i1} = -v_{i2} = 0.5\,\text{V}$, the output voltage is just the value of A_d.

Then for A_c:

Set

$$v_{i1} = v_{i2} = v_s = 1\,\text{V}$$

so that

$$v_d = v_{i1} - v_{i2}$$
$$= v_s - v_s$$
$$= 1\,\text{V} - 1\,\text{V}$$
$$= 0\,\text{V}$$

and

$$v_c = \frac{1}{2}(v_{i1} + v_{i2})$$
$$= \frac{1}{2}(1\,\text{V} + 1\,\text{V})$$
$$= 1\,\text{V}$$

Then the output voltage is

$$v_o = A_d v_d + A_c v_c$$
$$= A_d \cdot (0\,\text{V}) + A_c \cdot (1\,\text{V})$$
$$= A_c$$

So that it is clear that by setting the input voltages $v_{i1} = v_{i2} = 1\,\text{V}$, the output voltage is just the value of A_c.

After obtaining A_d and A_c, the value of common-mode rejection ratio (CMRR) can be calculated as follows:

$$\text{CMMR} = \frac{A_d}{A_c}$$

For engineers, the logarithmic form is more familiar [11, 16]

$$\text{CMMR(dB)} = 20\log_{10}\left(\frac{A_d}{A_c}\right)$$

As discussed before, the signal components of opposite polarity will be amplified greatly at the output, leading to large A_d values; the signal components that are in phase will mostly be canceled out, resulting in very small A_c values.

Ideally, the value of the CMRR is infinite. However, in practice, it is a very large number, and the larger the number, the better the op-amp performance will be.

Example 5.1

The input and output parameters of the op-amp are given Fig. 5.11. Calculate the value of the CMRR [11, 16].

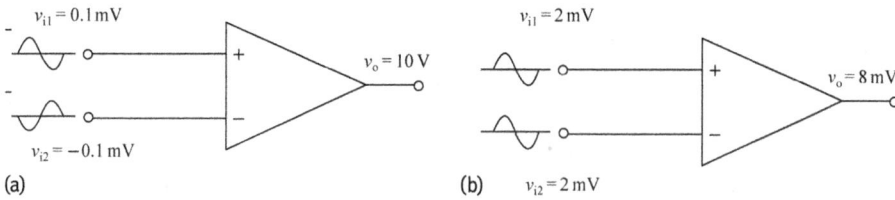

Fig. 5.11: Network of Example 5.1, (a) Differential input, (b) Common-mode.

Solution

From Fig. 5.11 (a), we obtained

$$v_d = v_{i1} - v_{i2}$$
$$= 0.5\,\text{mV} - (-0.5\,\text{mV})$$
$$= 1\,\text{mV}$$

In the differential input mode,

$$v_o = A_d v_d$$

So that

$$A_d = \frac{v_o}{v_d}$$
$$= \frac{10\,\text{V}}{1\,\text{mV}}$$
$$= 10^4$$

From Fig. 5.11 (b), we obtain

$$v_c = \frac{1}{2}(v_{i1} + v_{i2})$$
$$= \frac{1}{2}(2\,\text{mV} + 2\,\text{mV})$$
$$= 2\,\text{mV}$$

In in the common-mode input operation,

$$v_o = A_c v_c$$

So that

$$A_c = \frac{v_o}{v_c}$$
$$= \frac{8\,\text{mV}}{2\,\text{mV}}$$
$$= 4$$

So that

$$\text{CMMR} = \frac{A_d}{A_c}$$
$$= \frac{10^4}{4}$$
$$= 2.5 \times 10^3$$

or in decibel form,

$$\text{CMMR(dB)} = 20 \log_{10}\left(\frac{A_d}{A_c}\right)$$
$$= 20 \log_{10}(2.5 \times 10^3)$$
$$= 20 \times (3 + \log_{10}(2.5))$$
$$= 20 \times (3 + 0.398)$$
$$= 67.96 \, \text{dB}$$

5.4 Op-amp basic operations

Until now, fundamental knowledge about op-amp was presented. In this section, the properties and operation of op-amps will be further discussed from the viewpoint of engineering.

5.4.1 Transfer characteristic

An operational amplifier (op-amp) is an amplifier with a very high differential gain, featuring in very high input impedance (typically a few MΩ) and low output impedance (normally tens of Ω). The two inputs indicated by plus and minus signs can result in in-phase and out-of-phase output signals respectively. So, from the viewpoint of application, they are called noninverting and inverting input terminals, as shown in Fig. 5.12 [1, 11, 16].

Note that in Fig. 5.12, the inverting input terminal is drawn in the upper part of the symbol for the convenience of circuit design, which is important for the op-amp function and will be discussed below [1, 16].

The relationship between output voltage v_o and differential input v_d is described by the op-amp transfer characteristic, as shown in Fig. 5.13 [11, 16].

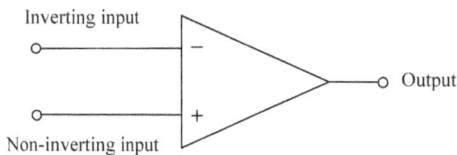

Fig. 5.12: Op-amp symbols indicating its properties.

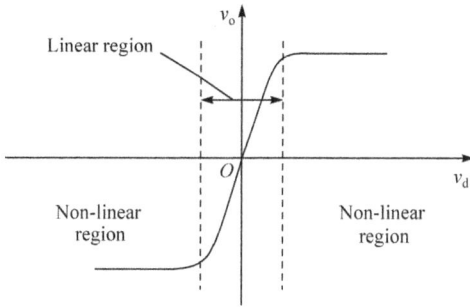

Fig. 5.13: Op-amp transfer characteristic.

As shown in Fig. 5.13, the transfer characteristic has two main parts, the linear region, which is located between the dashed parallel lines, and nonlinear regions, which are located outside the dashed parallel lines. The AC equivalent circuit when op-amp is working in the linear region, which has a finite increasing slope, is shown in Fig. 5.14. It can be seen that the input differential signal is applied between the two input terminals with an input impedance R_i that is typically very high. In the output port, the controlled voltage source has a voltage of the product of the amplifier differential gain and the input differential voltage. The output voltage is extracted from the controlled voltage source through output impedance R_o, which is typically very low.

The ideal AC equivalent circuit of the op-amp is shown in Fig. 5.15, which has an infinite input impedance R_i and a zero output impedance R_o.

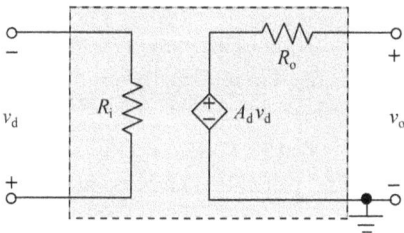

Fig. 5.14: Practical op-amp AC equivalent circuit when in the linear region.

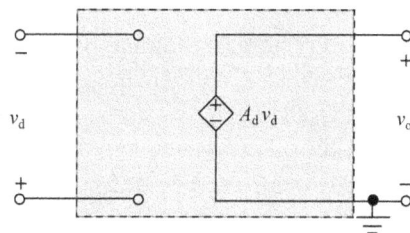

Fig. 5.15: Ideal op-amp AC equivalent circuit when in the linear region.

Note that the ideal op-amp AC equivalent circuit corresponds to the transfer characteristic with an infinitely increasing slope, or a vertical straight line. This is actually unstable for practical circuits and some measurements could be taken to attenuate its amplification ratio to make the circuit stable.

5.4.2 Basic operations

The basic op-amp network, which acts as a constant-gain amplifier, is shown in Fig. 5.16. An input signal v_i is applied through resistor R_1 to the inverting input terminal. More importantly, there is an additional resistor R_f connecting the inverting input and output terminals, the function of which will be discussed in the following.

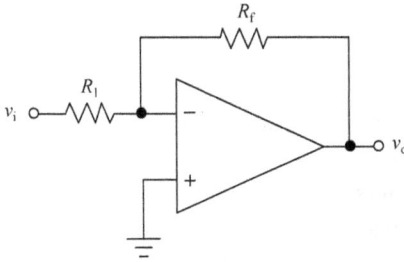

Fig. 5.16: Basic op-amp network.

The noninverting input terminal is connected to the ground. Due to the fact that v_i is essentially fed to the inverting input terminal, the consequent output is reversed in phase to the input signal. In other words, it is an inverting amplifier [1, 11, 16].

To carry out the analysis of the basic op-amp network, the AC equivalent circuit is used to replace the op-amp, as shown in Fig. 5.17. Note that the differential input voltage has its positive terminal at the bottom, because the noninverting input terminal is at the bottom.

To further simplify the analysis, the ideal AC equivalent circuit, which has infinite input impedance R_i and zero output impedance R_o, is used to replace the op-amp, as shown in Fig. 5.18.

Now, the objective is to find the relationship between v_o and v_i. As in Fig. 5.18, there are two sources, $A_d v_d$ and v_i, so the superposition principle can be used to find first the voltage v_d.

When only $A_d v_d$ exists, that is, by setting v_i to zero, the equivalent circuit is illustrated as in Fig. 5.19.

The controlled voltage source should now be regarded as an independent voltage source with a voltage output of $A_d v_d$. The network looks like the structure of voltage divider, so,

$$v_{d1} = -\frac{R_1}{R_1 + R_f} A_d v_d$$

When only v_i exists, that is, by setting $A_d v_d$ to zero, the equivalent circuit is as illustrated in Fig. 5.20. This also looks like the network of the voltage divider, so,

$$v_{d2} = -\frac{R_f}{R_1 + R_f} v_i$$

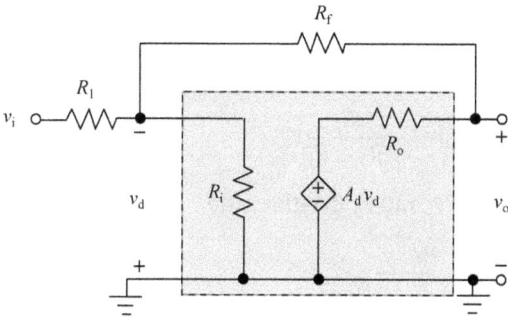

Fig. 5.17: Basic op-amp network with op-amp AC equivalent circuit.

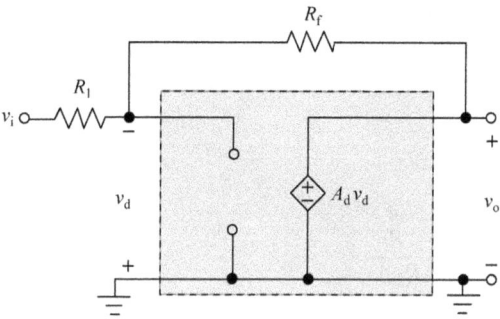

Fig. 5.18: Basic op-amp network with ideal op-amp AC equivalent circuit.

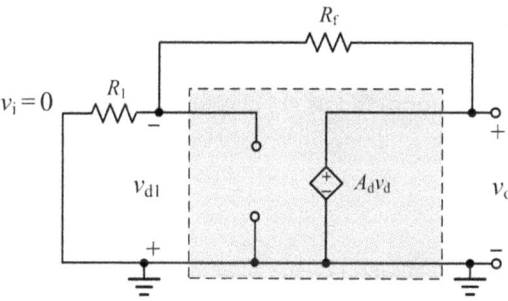

Fig. 5.19: Use of the superposition principle: with $A_d v_d$ only.

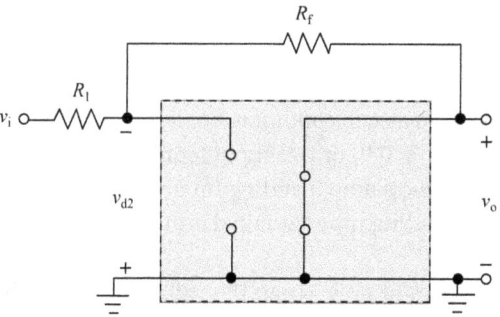

Fig. 5.20: Use of the superposition principle: with v_i only.

So, combining each component, we obtain

$$v_d = v_{d1} + v_{d2}$$

$$= -\frac{R_1}{R_1 + R_f}A_d v_d - \frac{R_f}{R_1 + R_f}v_i$$

Rearranging the equation and moving v_i and v_d to either side,

$$v_d = -\frac{R_f}{R_1(1 + A_d) + R_f}v_i$$

Normally, A_d is very large, so $R_1(1 + A_d) \gg R_f$, so that

$$v_d = -\frac{R_f}{R_1 A_d}v_i$$

Also, from Fig. 5.18, it is clear that

$$v_o = A_d v_d$$

Then

$$v_d = \frac{v_o}{A_d} = -\frac{R_f}{R_1 A_d}v_i$$

That is,

$$\frac{v_o}{v_i} = -\frac{R_f}{R_1} \tag{5.4}$$

This shows that the overall voltage gain of the op-amp network is only dependent on the values of resistors R_1 and R_f, and has nothing to do with A_d. This is an important conclusion based on which other linear op-amp circuits are developed.

5.4.3 Golden rules of op-amps

Normally, the output voltage v_o peak value is limited by the power supply of op-amp, which is mostly several volts. As discussed before, the voltage gain of op-amp A_d is very high, so the input differential v_d has the following equation:

$$v_d = \frac{v_o}{A_d}$$

This means that v_d could be very small in a such condition. So, it is reasonable to assume it to be nearly zero. The fact that $v_d \approx 0\,\text{V}$, or two input terminals are at the same level, results in the concept that at the op-amp inverting input terminal, there exists a virtual ground, whereas the noninverting input terminal is at real ground, as illustrated in Fig. 5.21.

Furthermore, from the viewpoint of current flow, when the voltage v_d is nearly zero, the two input terminals have the same voltage level, that is, a virtual short circuit

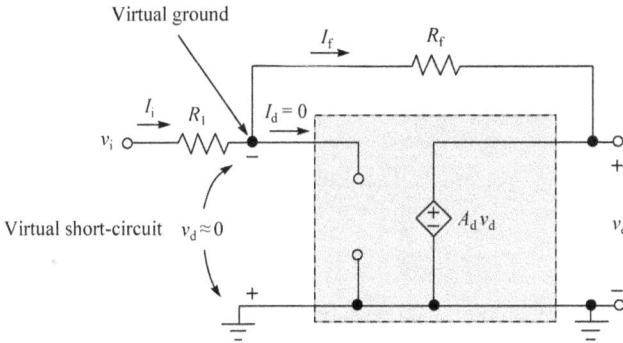

Fig. 5.21: Virtual ground and virtual short circuit of the op-amp.

exists between them. As there is no current flowing between them, the short circuit is not real but a virtual one, as illustrated in Fig. 5.21 [1, 11, 16].

The concepts of virtual ground and virtual short circuit are sometimes called the golden rules of op-amp, which indicates their usefulness in the analysis and applications of op-amps.

Based on the idea of the golden rules of op-amp, the voltage across R_1 is just the input voltage v_i, so the currents through R_1 can be obtained as

$$I_i = \frac{v_i}{R_1}$$

and the currents through R_f as

$$I_f = -\frac{v_o}{R_f}$$

Also, it is clear that $I_i = I_f$, due to $I_d = 0$. So,

$$\frac{v_i}{R_1} = -\frac{v_o}{R_f}$$

Then,

$$\frac{v_o}{v_i} = -\frac{R_f}{R_1}$$

This is the conclusion obtained before, as in Eq. (5.4).

Note that the golden rules depend on a large value of A_d and the existence of the feedback resistor R_f, resulting in a straightforward solution to determine the overall voltage gain v_o/v_i. Although the conclusion is based on some approximation, it really is a convenient way to obtain the overall voltage gain v_o/v_i.

5.4.4 Basic op-amp circuits

After the introduction of the basic op-amp network of Fig. 5.16, modifications can be carried out to meet some requirements leading to some special functional circuits.

These simple op-amp networks can serve as building blocks to many other more complicated networks [1, 11, 16].

1. Unity gain circuit

Based on the fundamental op-amp network of Fig. 5.16, if $R_1 = R_f = R$, as shown in Fig. 5.22, then

$$\frac{v_o}{v_i} = -\frac{R_f}{R_1} = -\frac{R}{R} = -1$$

So that the circuit presents a unity voltage gain with reversed phase. In practice, the accuracy of the values of these resistors will determine the real value of the voltage gain.

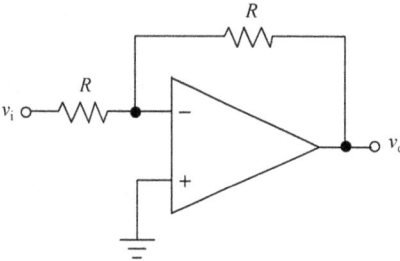

Fig. 5.22: Unity gain circuit.

2. Constant gain circuit

Also based on the fundamental op-amp network of Fig. 5.16, if $R_1 = R$ and $R_f = kR$, where k is a constant, then

$$\frac{v_o}{v_i} = -\frac{R_f}{R_1} = -\frac{kR}{R} = -k$$

So that the circuit presents a constant voltage gain k with reversed phase. In Fig. 5.23, the resistor $R_1 = 4.7\,k\Omega$ and $R_f = 47\,k\Omega$, and then $k = 10$. So the output voltage has a magnitude that is 10 times larger than that of the input with 180° phase shift.

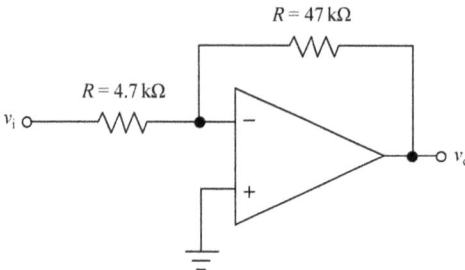

Fig. 5.23: Constant gain circuit.

In practice, the accuracy of the values of these resistors will determine the real value of the voltage gain. Sometimes, the surrounding temperature will affect the values of resistors.

3. Inverting amplifiers

The fundamental op-amp network of Fig. 5.16 is mostly called an inverting amplifier and is redrawn in Fig. 5.24. The output is the amplification of the input with a factor determined by the input resistor R_1 and the feedback resistor R_f.

$$v_o = -\frac{R_f}{R_1}v_i$$

The output has a 180° phase shift with respect to the input.

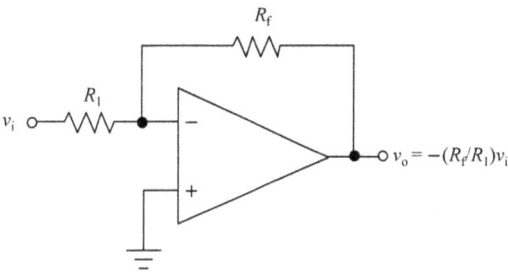

Fig. 5.24: Inverting amplifier.

4. Noninverting amplifier

Opposite to the inverting amplifier, there is also another op-amp network that can provide an amplification of the input voltage with the same phase, the noninverting op-amp amplifier, as shown in Fig. 5.25.

The structure of the noninverting op-amp amplifier has some similarity to the inverting op-amp amplifier in that there is a feedback resistor R_f between the inverting input terminal and the output terminal. Also, there is a resistor R_1 connecting to both feedback resistor R_f and the inverting input terminal. However, the difference is obvious. The other end of the resistor R_1 is grounded, instead of connecting to the input signal v_i. Actually, the input signal v_i is applied to the noninverting input terminal (positive terminal) directly.

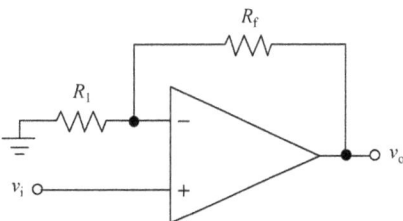

Fig. 5.25: Noninverting amplifier.

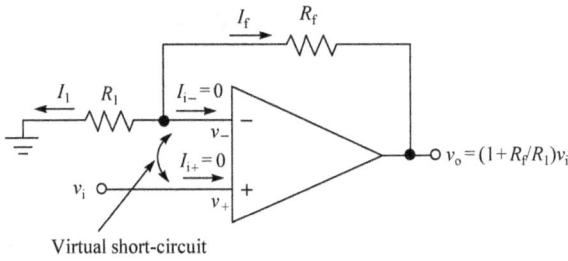

Fig. 5.26: Analysis of the noninverting amplifier.

It is obvious that the op-amp is an ideal one with large A_d and a feedback resistor R_f exists. So, the golden rules of op-amp are applicable to analyzing the noninverting op-amp amplifier, as shown in Fig. 5.26. Then, the virtual short circuit exits between the two input terminals. Thus, the voltage levels in the two input terminals are the same, that is,

$$v_+ = v_- = v_i$$

Then, the currents through R_1 can be obtained

$$I_1 = \frac{v_-}{R_1}$$
$$= \frac{v_i}{R_1}$$

The current through R_f is

$$I_f = \frac{v_- - v_o}{R_f}$$
$$= \frac{v_i - v_o}{R_f}$$

It is obvious that $I_- = 0$ A, so,

$$I_1 = -I_f$$

and

$$\frac{v_i}{R_1} = -\frac{v_i - v_o}{R_f}$$

Then,

$$v_i\left(\frac{1}{R_1} + \frac{1}{R_f}\right) = \frac{v_o}{R_f}$$

and

$$\frac{v_o}{v_i} = R_f\left(\frac{1}{R_1} + \frac{1}{R_f}\right)$$

That is,

$$\frac{v_o}{v_i} = \left(1 + \frac{R_f}{R_1}\right)$$

So, the output voltage is the product of the input voltage and a constant determined by R_1 and R_f, with the same phase as that of the input.

5. Unity follower

The unity-follower circuit, shown in Fig. 5.27, presents a unity gain with no phase shift or polarity reversal. In other words, the output signal is just a copy of the input one.

The same as for the noninverting op-amp amplifier, the op-amp is an ideal one with large A_d and the entire v_o acts as a feedback voltage. So, the golden rules of op-amp are applicable for analyzing the unity-follower circuit. Then, the virtual short circuit exists between the two input terminals, as shown in Fig. 5.28.

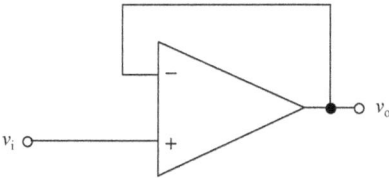

Fig. 5.27: Unity follower.   **Fig. 5.28:** Analysis of the unity follower.

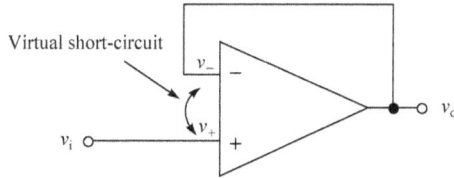

Thus, the voltage levels in the two input terminals are the same, that is,

$$v_+ = v_- = v_i$$

Also, in the feedback loop, it is obvious that

$$v_- = v_o$$

So that

$$v_o = v_i$$

The unity-follower circuit has the same role as an emitter follower in Section 3.4.7 or a source follower in Section 4.7.4, except that the gain is exactly unity.

5.5 Linear applications of op-amps

An operational amplifier is a differential amplifier characterized by very high gain, high input impedance and low output impedance. Typical ways to use the op-amp include but are not limited to the following: voltage amplitude modifications, sine wave or pulse oscillators, functional transformations of signals, active filters, linear operations of scaling, summing, subtracting and integrating, nonlinear operations of limiting, logarithmation, rectification of alternating voltages and instrumentation amplifiers. In the following section, some typical op-amp application circuits will be covered [1, 11, 16].

5.5.1 Voltage summing

The voltage summing circuit is one of the most widely used op-amp networks in the analog circuit world. It is the direct extension of the sum operation in mathematics to electronics. The circuit is shown in Fig. 5.29.

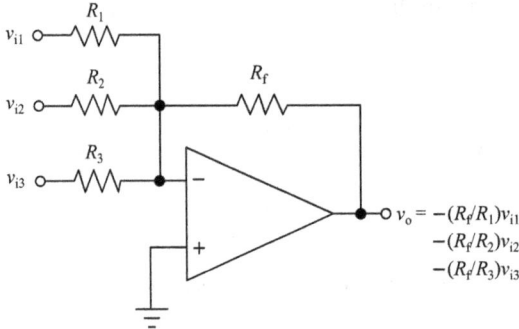

Fig. 5.29: Voltage summing circuit.

The same as for the inverting op-amp amplifier, the op-amp is assumed as an ideal one with large A_d and a feedback resistor R_f exists. So, the golden rules of op-amp are applicable for analyzing the voltage summing circuit. The first step to the analysis is using the superposition theorem to determine three output components one by one. As shown in Fig. 5.30, keep the jth input component v_{ij} (j = 1, 2, 3) as the original and set the others as zero, so the output is the corresponding component of v_{oj} (j = 1, 2, 3).

This is just what was discussed before, the inverting amplifier. So the output component is

$$v_{oj} = -\frac{R_f}{R_j}v_{ij}, \quad j = 1, 2, 3$$

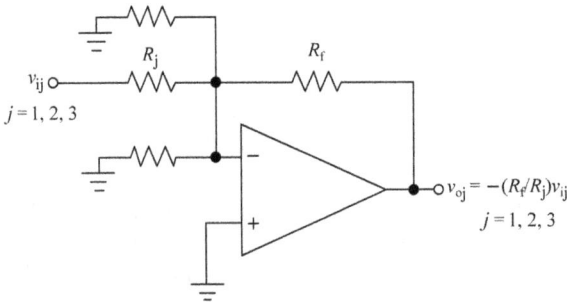

Fig. 5.30: Superposition theorem to analyze the voltage summing circuit.

Then, the overall output is

$$v_o = \sum_{j=1}^{3} v_{oj}$$

$$= \sum_{j=1}^{3} -\frac{R_f}{R_j} v_{ij}$$

$$= -\frac{R_f}{R_1} v_{i1} - \frac{R_f}{R_2} v_{i2} - \frac{R_f}{R_3} v_{i3}$$

Note that the output is actually a negative summation of all input components.

5.5.2 Voltage subtraction

Opposite to voltage summing, the voltage subtraction circuit is also common. It can perform the substrate operation between input signals. First, a single-stage circuit of voltage summing will be discussed (Fig. 5.31).

This is actually a combination of inverting and noninverting op-amp circuits. Since there are two input signals, the superposition theorem will be used.

Set v_{i2} to zero and keep v_{i1} unchanged; it is actually an inverting amplifier, as shown in Fig. 5.32.

So the relationship between v_{o1} and v_{i1} is

$$v_{o1} = -\frac{R_f}{R_1} v_{i1}$$

Then setting v_{i1} to zero and keeping v_{i2} unchanged; this is actually a noninverting amplifier, as shown in Fig. 5.33. The same as before, the op-amp is assumed an ideal one with large A_d and a feedback resistor R_f exists. So the golden rules of op-amp are applicable. Then, the virtual short circuit exits between the two input terminals, as shown in Fig. 5.33.

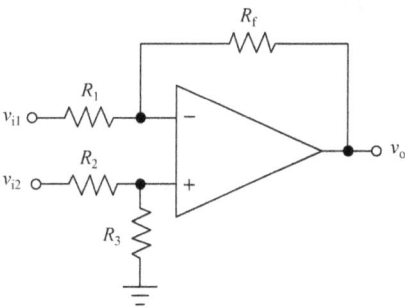

Fig. 5.31: Single-stage voltage subtraction circuit.

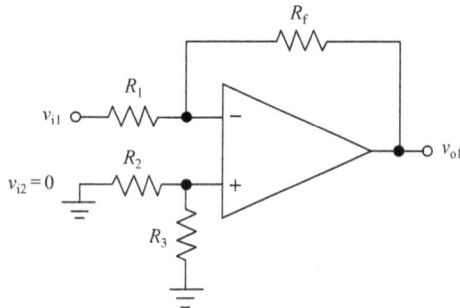

Fig. 5.32: Analysis of single-stage voltage subtraction circuit with v_{i2} set to zero.

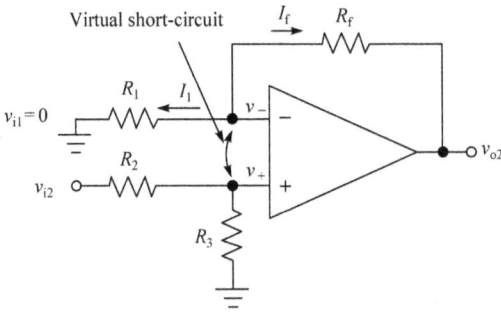

Fig. 5.33: Analysis of the single-stage voltage subtraction circuit with v_{i1} set to zero.

Thus, the voltage levels in the two input terminals are same, that is,

$$v_+ = v_-$$

Moreover,

$$v_+ = \frac{R_3}{R_2 + R_3} v_{i2}$$

Then, the currents through R_1 can be obtained,

$$I_1 = \frac{v_-}{R_1}$$
$$= \frac{v_+}{R_1}$$
$$= \frac{R_3}{R_1(R_2 + R_3)} v_{i2}$$

The current through R_f is

$$I_f = \frac{v_- - v_{o2}}{R_f}$$
$$= \frac{v_+ - v_{o2}}{R_f}$$
$$= \frac{R_3}{R_f(R_2 + R_3)} v_{i2} - \frac{1}{R_f} v_{o2}$$

In fact, $I_1 = -I_f$, so,

$$-\frac{R_3}{R_1(R_2 + R_3)} v_{i2} = \frac{R_3}{R_f(R_2 + R_3)} v_{i2} - \frac{1}{R_f} v_{o2}$$

Moving the variables v_{i2} and v_{o2} to either side of the equation, we obtain

$$v_{o2} = \frac{R_3}{R_2 + R_3} \cdot \frac{R_1 + R_f}{R_1} v_{i2}$$

Then, combining the two output components,

$$v_o = v_{o1} + v_{o2}$$
$$= \frac{R_3}{R_2 + R_3} \cdot \frac{R_1 + R_f}{R_1} v_{i2} - \frac{R_f}{R_1} v_{i1}$$

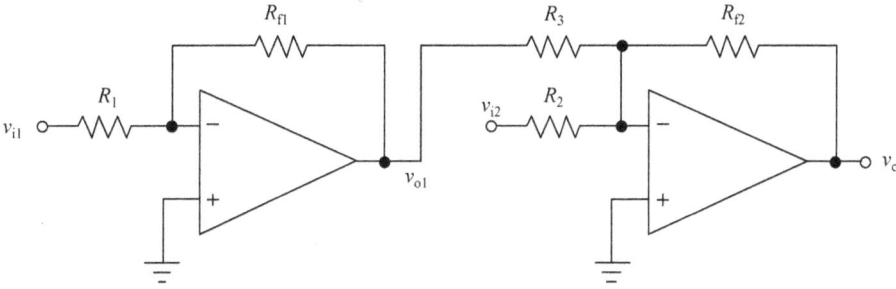

Fig. 5.34: Two-stage voltage subtraction circuit.

Note that suitable selection of the values of R_1, R_2, R_3 and R_f, will precisely control the output how you want.

Another structure of the voltage subtraction circuit is shown in Fig. 5.34.

This is actually a two-stage op-amp circuit, each stage of which is an inverting amplifier. So its analysis is once again the use of the superposition theorem. To facilitate the analysis, an auxiliary variable v_{o1} has been introduced in the output terminal of the first stage. So, the relationship between v_{i1} and v_{o1} is

$$v_{o1} = -\frac{R_{f1}}{R_1} v_{i1}$$

and the relationship between v_o and its inputs is

$$v_o = -\frac{R_{f2}}{R_3} v_{o1} - \frac{R_{f2}}{R_2} v_{i2}$$

Then, combining the above two relationships,

$$v_o = -\frac{R_{f2}}{R_3} \left(-\frac{R_{f1}}{R_1} v_{i1} \right) - \frac{R_{f2}}{R_2} v_{i2}$$
$$= \frac{R_{f2}}{R_3} \cdot \frac{R_{f1}}{R_1} v_{i1} - \frac{R_{f2}}{R_2} v_{i2}$$

By wisely selecting the values of R_1, R_2, R_3, R_{f1} and R_{f2}, the output can be obtained as what you need.

5.5.3 Integrator

All the op-amp circuits discussed so far have only used resistors as input and feedback components. However, to perform more complicated mathematical operations, other electric components should be involved in different positions for different purposes.

Based on the inverting amplifier, the feedback component is change to a capacitor, as shown in Fig. 5.35. This kind of op-amp circuit is normally called an integrator [22].

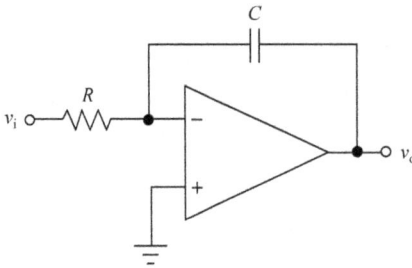

Fig. 5.35: Op-amp circuit of the integrator.

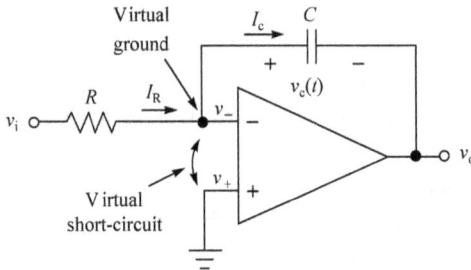

Fig. 5.36: Analysis of the integrator.

It is also assumed that the op-amp is an ideal one with large A_d and a feedback component C exists. So, the golden rules of op-amp are applicable for analyzing the integrator. Then, the virtual short circuit exists between the two input terminals and virtual ground exists in negative input terminal, as shown in Fig. 5.36.

Thus, the voltage levels in the two input terminals are same, that is,

$$v_+ = v_- = 0$$

Then, the currents through R can be obtained

$$I_R = \frac{v_i}{R}$$

For a capacitor C, from its own electrical property, it is obvious that

$$I_c = C\frac{dv_c(t)}{dt}$$

where C is capacitance of the capacitor, and $v_c(t)$ is AC voltage across the capacitor. In fact $I_R = I_c$, because no current enters into the op-amp through either input terminal, so,

$$\frac{v_i}{R} = C\frac{dv_c(t)}{dt}$$

Also, $v_o(t) = -v_c(t)$ with respect to the virtual ground, so

$$\frac{v_i}{R} = -C\frac{dv_o(t)}{dt}$$

Rearrangement of it leads to

$$\frac{dv_o(t)}{dt} = -\frac{1}{RC}v_i$$

In a more obvious form, the relations could be

$$v_o(t) = -\frac{1}{RC} \int v_i(t)dt$$

This shows that the output $v_o(t)$ is the integral of the input $v_i(t)$, with an inversion and scale factor of $1/RC$. It accumulates the input quantity $v_i(t)$ over a defined time to produce a representative output $v_o(t)$. This accounts for the name of the circuit.

The function of the integral of an input signal gives the analog computer the ability to solve differential equations. Integrators are the basis of analog computers and charge amplifiers. However, in recent decades integration has been performed mainly by digital computing algorithms.

Moreover, from other viewpoint, an integrator is a form of first-order low-pass filter, which can be performed in the continuous-time (analog) domain or approximated (simulated) in the discrete-time (digital) domain. An integrator will have a low pass filtering effect but when given an offset it will accumulate a value until it reaches a limit of the system or overflows.

For different input signals, integrators may have different types. A voltage integrator is an electronic device performing a time integration of an electric voltage, thus measuring the total volt-second product. A current integrator is an electronic device performing the time integration of an electric current, thus measuring a total electric charge. A charge amplifier is an example of a current integrator.

Also, an integrator can be combined with other op-amp circuits to create new circuits. For example, as shown in Fig. 5.37, it is the combination of voltage summing and an integrator.

Obviously, it has the following relationship:

$$v_o(t) = -\frac{1}{R_1 C} \int v_{i1}(t)dt - \frac{1}{R_2 C} \int v_{i2}(t)dt - \frac{1}{R_3 C} \int v_{i3}(t)dt$$

By wisely selecting the values of R_1, R_2, R_3 and C, the desired output can be obtained.

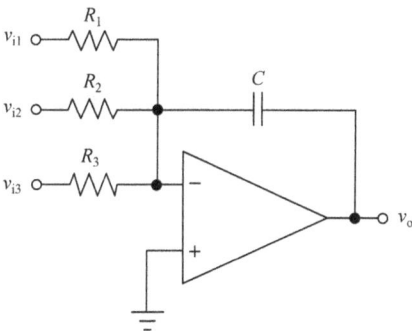

Fig. 5.37: Integrator with multiple inputs.

5.5.4 Differentiator

Based on an inverting amplifier, keeping the feedback resistor as before and changing the input resistor to a capacitor, a new kind of op-amp circuit, called a differentiator, results, as shown in Fig. 5.38.

It is also assumed that the op-amp is an ideal one with large A_d and that a feedback resistor R exists. So, the golden rules of op-amp are applicable for analyzing the differentiator. Then, the virtual short circuit exists between the two input terminals and virtual ground exists in the negative input terminal, as shown in Fig. 5.39 [23].

Thus, the voltage levels in the two input terminals are the same, that is,

$$v_+ = v_- = 0$$

For the input capacitor C, from its own electrical property, it is obvious that

$$I_c = C\frac{dv_c(t)}{dt}$$
$$= C\frac{dv_i(t)}{dt}$$

where C is the capacitance of the capacitor, and $v_c(t)$ is the AC voltage across the capacitor.

Then, the currents through R can be obtained

$$I_R = \frac{v_- - v_o}{R}$$
$$= -\frac{v_o}{R}$$

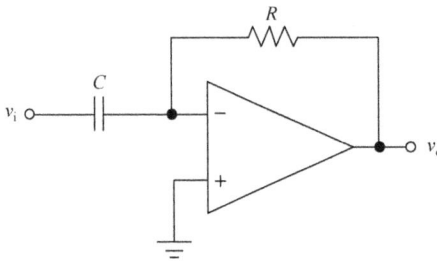

Fig. 5.38: Op-amp circuit of the differentiator.

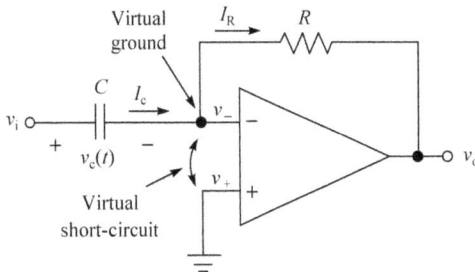

Fig. 5.39: Analysis of the differentiator.

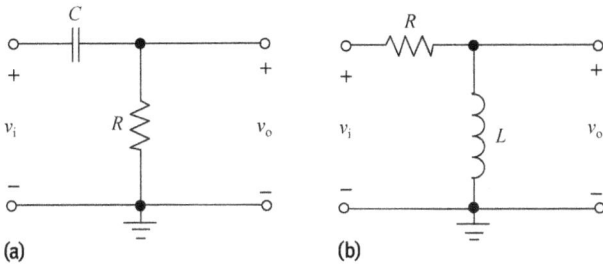

Fig. 5.40: Passive differentiators, (a) Capacitive differentiator, (b) Inductive differentiator.

In fact, $I_R = I_c$, because no current enters into the op-amp through either input terminal, so,

$$-\frac{v_o}{R} = C\frac{dv_i(t)}{dt}$$

that is,

$$v_o = -RC\frac{dv_i(t)}{dt}$$

This shows that the output $v_o(t)$ is the differentiation of the input $v_i(t)$, with an inversion and a scale factor of RC. This accounts for the name of the circuit.

The differentiator circuit is essentially a high-pass filter. It can generate a square wave from a triangle wave input and will produce alternating-direction voltage spikes when a square wave is applied. In ideal cases, a differentiator will reverse the effects of an integrator on a waveform, and vice versa. Differentiators are an important part of electronic analog computers.

A differentiator circuit made from an op-amp is sometimes called an active differentiator. On the other hand, passive differentiator circuits are basic electronic circuits that are widely used in circuit analysis based on the equivalent circuit method. Typical passive differentiator circuits are shown in Fig. 5.40.

5.5.5 Instrumentation amplifier

An instrumentation (or instrumentational) amplifier, as shown in Fig. 5.41 [24, 25], is a modified version of a differential amplifier that includes an additional input buffer stage, which eliminates the necessity of input impedance matching and thus makes it particularly suitable for equipment testing and instrument measurement. Additional features include very large input impedances, considerably high A_d, very high CMRR, low noise, low DC offset and low drift. It possesses the capability of controlling the overall gain of the amplifier circuit by adjusting a single resistor. Instrumentation amplifiers have great accuracy and high stability.

This complex circuit, as shown in Fig. 5.41, is constructed from a buffered differential amplifier stage, with resistors of R_G and two R_1s, linking the two buffer circuits

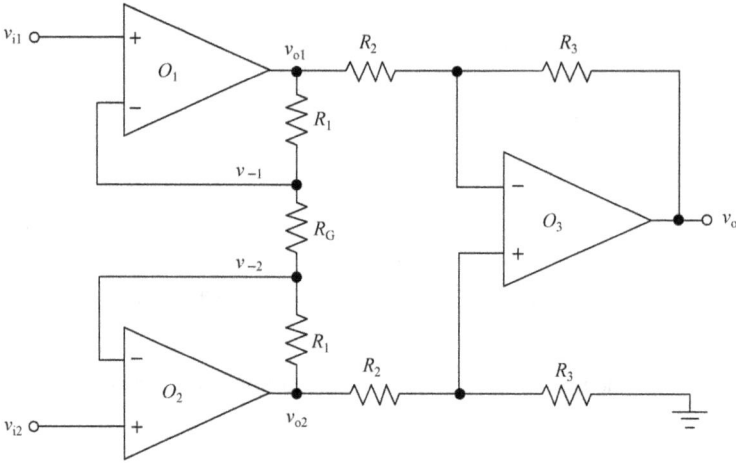

Fig. 5.41: Instrumentation amplifier.

together. Consider all resistors to be of symmetric values except for R_G. The negative feedback of op-amp O_1 causes the voltage v_{-1} to be equal to the input voltage of op-amp O_1, that is,

$$v_{-1} = v_{i1}$$

Likewise, the voltage v_{-2} is held at a value equal to the input voltage of op-amp O_2, that is,

$$v_{-2} = v_{i2}$$

This establishes a voltage drop across R_G, which is proportional to the voltage difference between v_{i1} and v_{i2}. This voltage drop causes a current through R_G, and due to the fact that the two input terminals of O_1 and O_2 draw no current, the current through R_G must go through the two R_2 resistors. This produces a voltage drop between v_{o1} and v_{o2} equal to

$$v_{o1} - v_{o2} = \left(1 + \frac{2R_1}{R_G}\right)(v_{i1} - v_{i2})$$

The regular differential amplifier of O_3 then takes this voltage drop of $(v_{o1} - v_{o2})$ and amplifies it with multiplier of R_3/R_2,

$$v_o = \frac{R_3}{R_2}(v_{o2} - v_{o1})$$

$$= \frac{R_3}{R_2}\left(1 + \frac{2R_1}{R_G}\right)(v_{i2} - v_{i1})$$

The single resistor R_G between the two inverting inputs of O_1 and O_2 can be used as a proper way to adjust the gain; it increases the differential-mode gain of O_1 and O_2, while leaving the common-mode gain equal to 1. This increases the CMRR of the circuit. Another benefit is that it boosts the gain using a single resistor, avoiding resistor-matching problem and conveniently adjusting the gain of the network by changing the value of a single resistor R_G [26].

In practice, a set of adjustable resistors or a potentiometer can be used for R_G to change the gain without the complexity of the resistor-matching problem.

5.5.6 Active filters with op-amps

The example of a passive filter is shown in Fig. 5.40. Actually, transfer functions can be implemented in either form of passive or active filters.

An active filter is a type of analog electronic filter that includes active components, one that has a power supply, such as an amplifier. This can improve the performance and predictability of the filter, while avoiding the need for inductors, which are normally costly compared to other components like resistors and capacitors. Also, an active filter with an op-amp can prevent the load impedance of the following stage from affecting the characteristics of the filter.

Without an inductor, an active filter can have complex poles and zeros. The tuned frequency, the shape of the response curve and the quality factor can often be manipulated with variable resistors. Normally, one parameter can be adjusted without affecting the others.

However, there are some limitations when using op-amps in filters. If basic filter design equations neglect the finite bandwidth of op-amps, the filters are often impractical at high frequencies. Moreover, op-amps consume power and introduce noise into the network. If a DC path is not provided for the bias current to the op-amp elements, circuit topologies may be impractical. Moreover, power consumption should be taken into consideration when an op-amp is involved.

Transfer functions can be implemented in the form of either active or passive filters. Transfer functions commonly include the following types [1, 3, 13, 27, 28]:

The low-pass filter, which provides a constant amplification from the DC component up to a cutoff frequency f_{OH} and attenuate all higher frequencies. The ideal response of the low-pass filter is shown in Fig. 5.42.

The high-pass filter, which attenuates from the DC component up to a cutoff frequency f_{OL} and provides a constant amplification above f_{OL}. The ideal response of the high-pass filter is shown in Fig. 5.43.

The bandpass filter, which provides a constant amplification between frequencies f_{OL} and f_{OH}, attenuates below the frequency f_{OL} and above the frequency f_{OH}. The ideal response of the bandpass filter is shown in Fig. 5.44.

The band-stop filter or band-rejection filter, which attenuates between frequencies f_{OL} and f_{OH}, provides a constant amplification below the frequency f_{OL} and above the frequency f_{OH}. The ideal response of the band- stop filter is shown in Fig. 5.45.

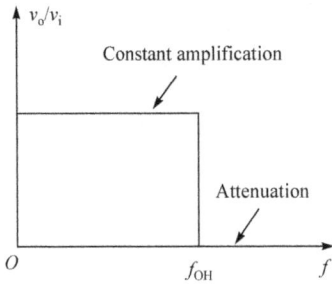

Fig. 5.42: Ideal response of the low-pass filter.

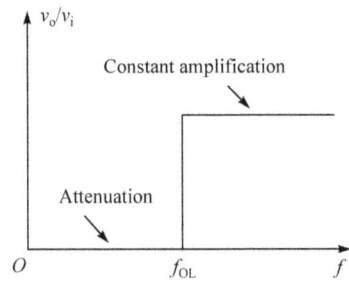

Fig. 5.43: Ideal response of the high-pass filter.

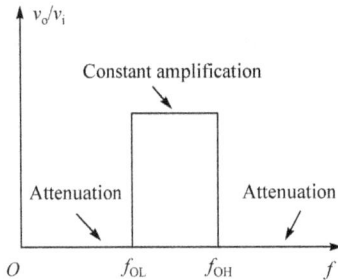

Fig. 5.44: Ideal response of the bandpass filter.

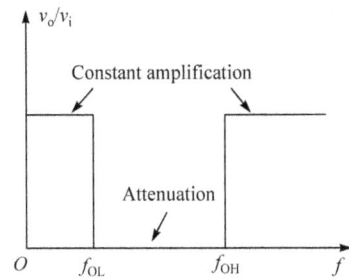

Fig. 5.45: Ideal response of the band-stop filter.

The band-stop filter is the opposite of the bandpass filter. A special type of band-stop filter is the notch filter, which attenuates certain frequencies while allowing all others to pass. Actually, it has a narrow stopband with a high Q factor.

Moreover, combinations of multiple filters are possible, such as notch filters and high-pass filters.

1. Low-pass filter [29]

Low-pass filters exist in many different forms, including electronic circuits, such as antialiasing filters for conditioning signals before analog-to-digital conversion, digital filters for smoothing sets of data, blurring of images to remove noise and so on.

Low-pass filters result in a smoother form of a signal, removing fast-changing fluctuations and keeping long-term slowly-changing components.

Moreover, the idea of low-pass filters can be used to generate the techniques for digital image processing in either the spatial or frequency domain, such as image unsharpening, blurring, de-noising and low-frequency component extraction. Figure 5.46 shows the results of a digital image filtered by a low-pass filter. It can be seen that only the low-frequency components of the image are remained.

One simple passive low-pass filter, as shown in Fig. 5.47, consists of a resistor in series with a capacitor. The input signal is applied to the resistor and output is extracted from the capacitor.

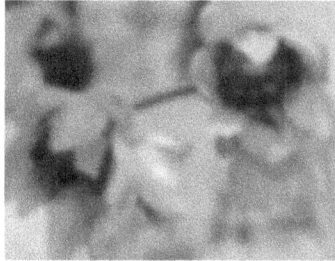

(a)   (b)

Fig. 5.46: Digital image processed by a low-pass filter, (a) Original image, (b) Processed by low-pass filter.

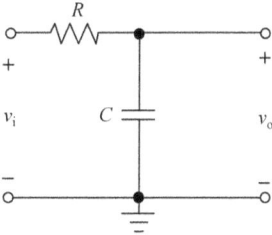

Fig. 5.47: Passive low-pass filter.

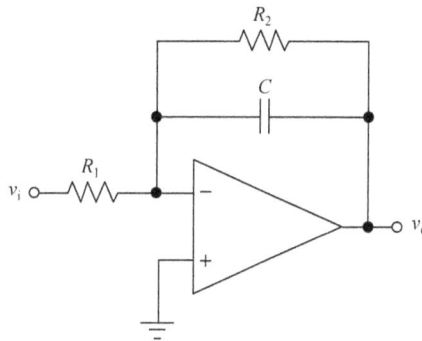

Fig. 5.48: An active first-order low-pass filter.

The capacitor exhibits reactance, blocking low-frequency signals from going through and forcing them out to the load instead. At higher frequencies, the reactance drops and the capacitor eventually acts as a short circuit, leading to passing through of higher frequencies. The cutoff frequency in hertz is determined by

$$f_{OH} = \frac{1}{2\pi RC}$$

An active first-order low-pass filter, implemented with an op-amp, resistors and capacitor is shown in Fig. 5.48.

Its cutoff frequency in hertz is defined as

$$f_{OH} = \frac{1}{2\pi R_2 C}$$

The voltage gain in the passband is $-R_2/R_1$, and the stopband drops off at -20 dB per decade, as illustrated in Fig. 5.49.

Another, a little more complicated, active first-order low-pass filter is shown in Fig. 5.50.

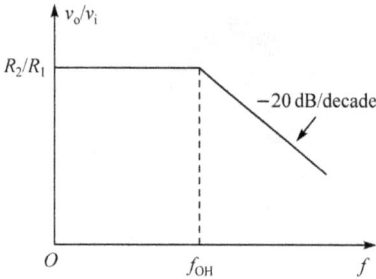

Fig. 5.49: Response of an active first-order low-pass filter.

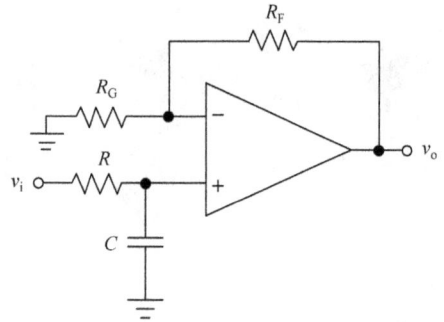

Fig. 5.50: Another active first-order low-pass filter.

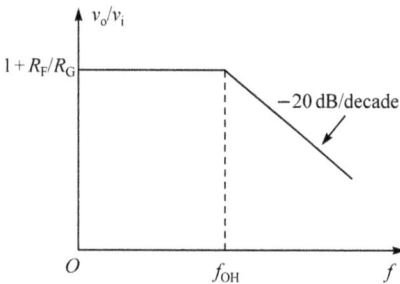

Fig. 5.51: Response of an active first-order low-pass filter.

Its cutoff frequency in hertz is defined as

$$f_{OH} = \frac{1}{2\pi RC}$$

The stopband drops off at −20 dB per decade, as illustrated in Fig. 5.51. The voltage gain in the passband is determined by resistors as

$$A_V = 1 + \frac{R_F}{R_G}$$

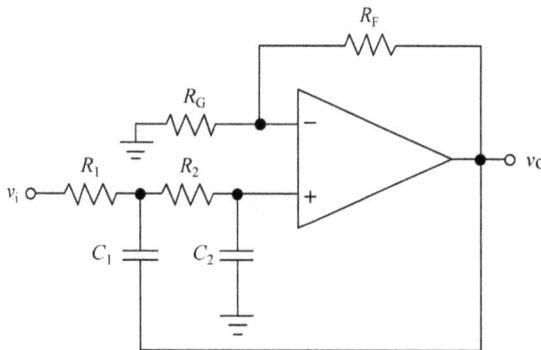

Fig. 5.52: A second-order active low-pass filter.

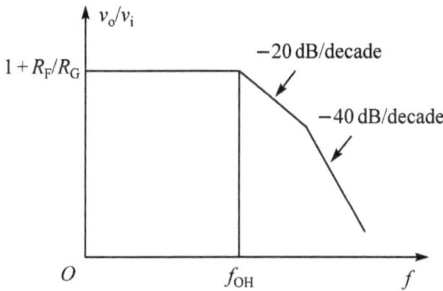

Fig. 5.53: Response of a second-order active low-pass filter.

Now, the shortcoming of response from the low-pass filter in Fig. 5.51 is obvious in that the attenuation in the stopband is too slow compared with the ideal response in Fig. 5.42. So, some modification can be made to get a better response.

A modified active low-pass filter is shown in Fig. 5.52. Actually, one additional passive low-pass filter of Fig. 5.47 has been introduced, leading to a second-order active low-pass filter.

The voltage gain in the passband is still determined as

$$A_V = 1 + \frac{R_F}{R_G}$$

Its cutoff frequency in hertz, which is the same as for the first-order one, is defined as

$$f_{OH} = \frac{1}{2\pi R_2 C_2}$$

However, the improvement is that the filter response declines at a quicker rate as a second-order filter, which is much closer to the ideal characteristic of Fig. 5.42, as shown in Fig. 5.53.

2. High-pass filter [30]

High-pass filters have uses in many fields. The central idea is to block DC components from circuitry that is sensitive to nonzero average voltages. For example, in audio applications, high-pass filters are used in an audio crossover to feed high frequencies to a tweeter while suppressing bass frequencies to avoid interference or damage to the speaker.

Also, for AC coupling at the inputs of audio power amplifiers, high-pass filters are used to avoid the amplification of DC currents which may lead to damage of the amplifier, loss of dynamic range for the amplifier and generation of unwanted heat at the loudspeaker's voice coil.

Moreover, for each audio channel strip in the mixing consoles, high-pass filtering could be fixed slope, fixed frequency, or sweepable within a specified frequency range. Further, the idea of high-pass filters can be used in digital image processing in either the spatial domain or the frequency domain, such as image sharpening, high-boost filtering and edge detection. Figure 5.54 shows the results of a digital image processed by a high-pass filter, indicating the remained fast-changing regions in the image, that is, edges. This can be further processed to improve the visual effect of the image.

(a)   (b)

Fig. 5.54: Digital image processed by a high-pass filter, (a) Original image, (b) Processed by high-pass filter.

However, the limitation of high-pass filters for image processing is that they may increase the noise level of the image. Generally, high-pass filters can also be used in conjunction with a low-pass filter to produce a bandpass filter.

One simple passive first-order high-pass filter, as shown in Fig. 5.55, consists of a capacitor in series with a resistor. The input signal is applied to the capacitor and output is extracted from the resistor. The capacitor can block low-frequency signals with its high reactance and let high-frequency components go through it and generate output voltage on the resistor. The cutoff frequency in hertz is determined by

$$f_{\text{OL}} = \frac{1}{2\pi RC}$$

An active first-order high-pass filter, implemented with an op-amp, resistors and capacitor, is shown in Fig. 5.56. Its cutoff frequency in hertz is defined as

$$f_{\text{OL}} = \frac{1}{2\pi R_1 C}$$

The voltage gain in the passband is $-R_2/R_1$, and the stopband increases at 20 dB per decade, as illustrated in Fig. 5.57.

Another form of an active first-order high-pass filter is shown in Fig. 5.58. Its cutoff frequency in hertz is defined as

$$f_{\text{OL}} = \frac{1}{2\pi RC}$$

The stopband increases at 20 dB per decade, as illustrated in Fig. 5.59. The voltage gain in the passband is determined by resistors as

$$A_{\text{v}} = 1 + \frac{R_{\text{F}}}{R_{\text{G}}}$$

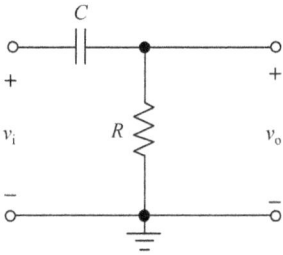

Fig. 5.55: Passive first-order high-pass filter.

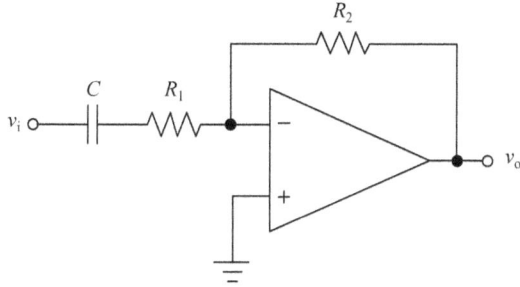

Fig. 5.56: Active first-order high-pass filter.

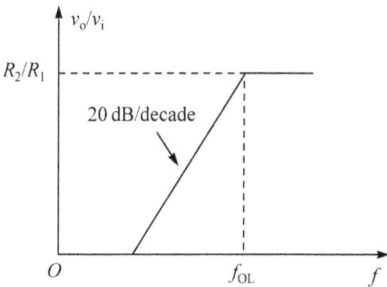

Fig. 5.57: Response of an active first-order high-pass filter.

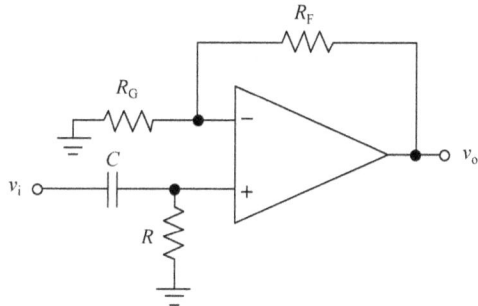

Fig. 5.58: Another form of active first-order high-pass filter.

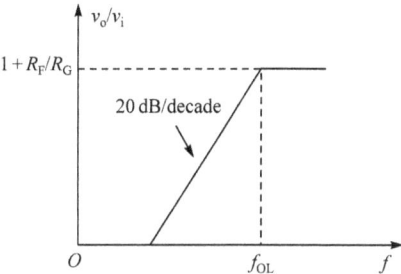

Fig. 5.59: Response of an active first-order high-pass filter.

Moreover, some improvement can be made to get a steeplier increasing slope.

A modified active low-pass filter is shown in Fig. 5.60. Actually, one additional passive high-pass filter of Fig. 5.55 has been introduced, leading to a second-order active low-pass filter.

The voltage gain in the passband is still determined as

$$A_V = 1 + \frac{R_F}{R_G}$$

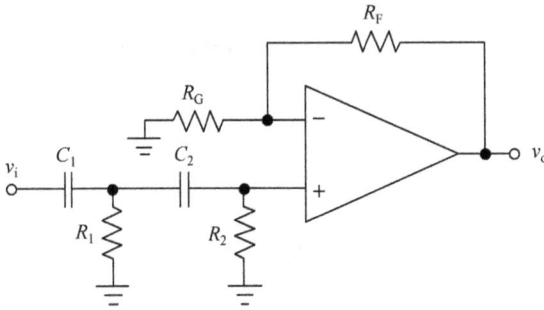

Fig. 5.60: A second-order active high-pass filter.

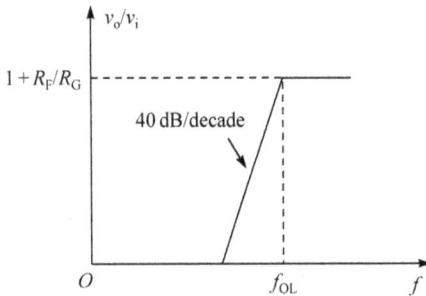

Fig. 5.61: Response of a second-order active high-pass filter.

Its cutoff frequency in hertz is defined as

$$f_{OL} = \frac{1}{2\pi R_2 C_2}$$

The improvement is obvious: the filter response rises at a quicker rate as a second-order filter, which is much closer to the ideal characteristic of Fig. 5.43, as shown in Fig. 5.61.

3. Bandpass filter [31]

Bandpass filters are commonly used in transmitters and receivers of telecommunications. In a transmitter, a bandpass filter is to limit the bandwidth of the output signal to the allocated band, avoiding interference to other stations. In a receiver, a bandpass filter allows signals within a selected spectrum to be processed, avoiding the reception of signals at unwanted frequencies.

For both transmitter and receiver, well-designed bandpass filters can lead to bandwidth optimization, maximization of number of signal transmitters and minimization of interference or competition among signals. An example of an analog electronic bandpass filter is the RLC circuit shown in Fig. 5.62.

The bandpass filter has two important parameters: the bandwidth and the quality factor (Q factor). The bandwidth is measured between the two half-power points, which are determined by gain of -3 dB, or half of the square root of 2, or approximately 0.707, with respect to the peak value on the curve of the transfer function of the bandpass filter, as shown in Fig. 5.63.

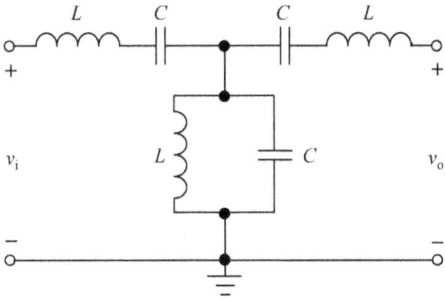

Fig. 5.62: Passive bandpass filter with RLC.

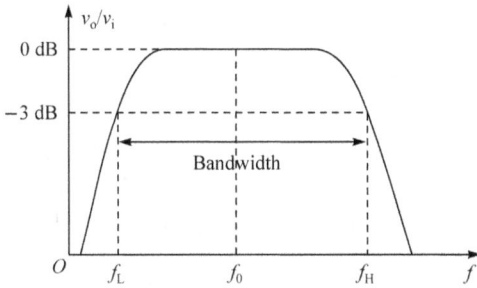

Fig. 5.63: Bandwidth of a bandpass filter.

The quality factor (Q factor) is also a character of a bandpass filter. It is defined as the reciprocal of the fractional bandwidth. A high-Q filter will have a narrow pass-band, which is normally called the narrow-band filter. A low-Q filter will have a wide passband, called the wide-band filter.

An active bandpass filter can be set up by cascading a low-pass filter with a high-pass filter, both of which were discussed earlier. An example is shown in Fig. 5.64.

The voltage gain in the passband is determined by that of each stage as

$$A_\mathrm{v} = \left(1 + \frac{R_{F1}}{R_{G1}} \right)\left(1 + \frac{R_{F2}}{R_{G2}} \right)$$

and the transfer function can be illustrated as in Fig. 5.65.

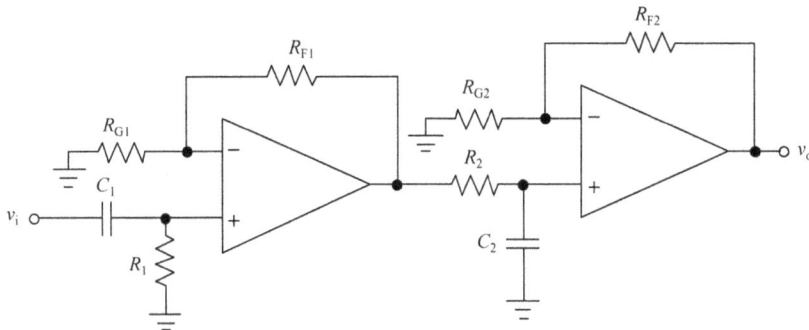

Fig. 5.64: An active bandpass filter with two op-amps.

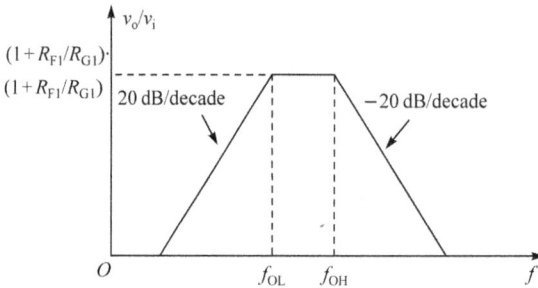

Fig. 5.65: Response of an active bandpass filter.

5.6 Nonlinear applications of op-amps

When the op-amp is involved in the circuits to generate an output signal with the same waveform as that of the input, except that the output has a larger magnitude, the circuit belongs to a linear circuit. On the other hand, if the output signal has a different waveform to the input, the circuit is a nonlinear one.

Also, from the viewpoint of signal frequency, if the output signal contains the same frequency components as those of the input signal, the circuit is a linear one; otherwise, it is nonlinear.

The op-amp amplifiers with negative feedback, discussed so for, are mostly linear circuits. In the following, some typical nonlinear op-amp applications will be discussed.

5.6.1 Comparators

In electronics, a comparator is a circuit that compares two input voltages or currents and outputs a digital signal indicating which one is larger. It has two analog input terminals v_+, v_- and one binary digital output v_0 [11, 13, 16]. The ideal output is

$$v_0 = \begin{cases} 1, & \text{if } v_+ > v_- \\ 0, & \text{if } v_+ < v_- \end{cases}$$

where the output 0 and 1 are logic levels, corresponding to predetermined voltages. A comparator can be implemented by a high-gain differential amplifier in an open loop (without negative feedback) with bipolar supply, as shown in Fig. 5.66.

Rail-to-rail comparators allow any differential voltages within the power supply range. When powered from a bipolar (dual rail) supply,

$$V_- \le v_+, \quad v_- \le V_+$$

On the other hand, when powered from a unipolar power supply for TTL/CMOS chips:

$$0 \le v_+, \quad v_- \le V_{CC}$$

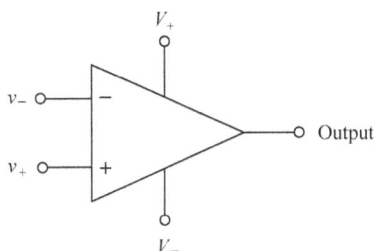

Fig. 5.66: An op-amp used as comparator.

An op-amp is normally characterized by a well-balanced difference input and very high gain. This parallels the characteristics of comparators and can be used in applications with low-performance requirements.

In theory, without negative feedback, a standard op-amp operating in open-loop configuration may be used as a comparator. When the noninverting input v_+ is at a higher voltage than the inverting input v_-, the high gain of the op-amp causes the output to saturate at the logic 1 voltage (near V_+). When the noninverting input v_+ drops below the inverting input v_-, the output saturates at the logic 0 voltage (near V_-).

The ideal transfer characters operating in nonlinear mode are shown in Fig. 5.67 (a). In practice, the transfer characters will not have a vertical rise, as in Fig. 5.67 (b).

Normally, a reference voltage v_{ref} is applied to the inverting input terminal and real input signal v_+ is fed to the noninverting terminal, as shown in Fig. 5.68.

As long as input v_+ is higher than v_{ref}, the output is logic 1 (near V_+). When input v_+ drops below v_{ref}, the output is logic 0 (near V_-). In Fig. 5.69, a small positive

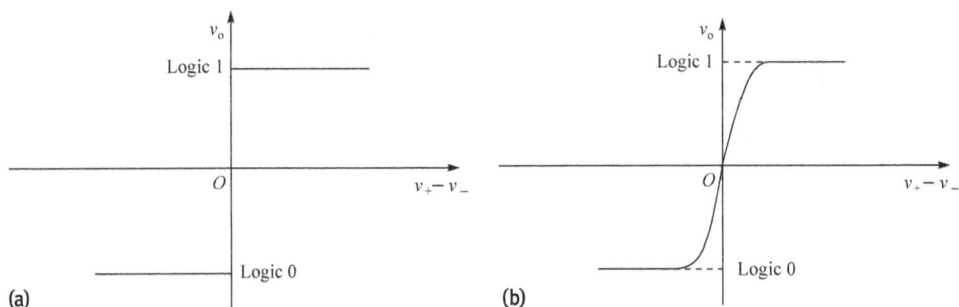

(a) (b)

Fig. 5.67: Transfer characteristics of a comparator, (a) Ideal, (b) Practical.

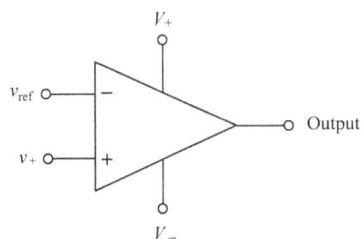

Fig. 5.68: Reference voltage is used in a comparator.

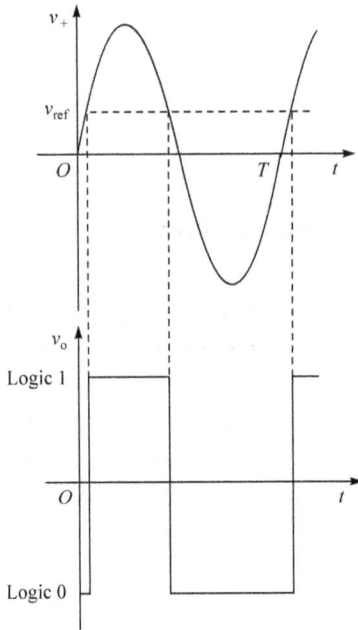

Fig. 5.69: Input, reference and output waveforms of a comparator.

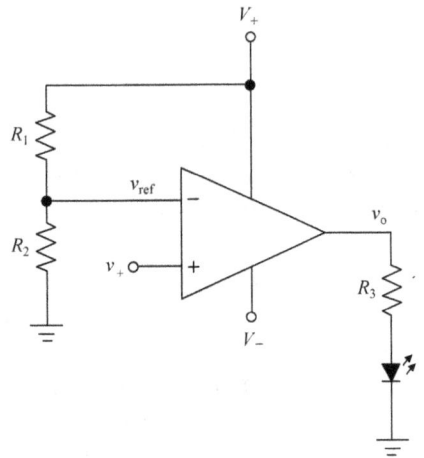

Fig. 5.70: LED connected to the output of a comparator.

voltage is selected as the reference voltage applied to the inverting input terminal. The input signal v_+ is a sinusoid waveform. The output is a rectangular waveform with the corresponding width determined by the intersections of input and reference signals. Generally, the reference signal could be positive, negative and zero. It can also be applied to either inverting or noninverting input terminals.

For a practical purposes, the output of a comparator could be connected with an LED to show the output logic visually, as shown Fig. 5.70. The reference voltage is determined by R_1 and R_2. The resistor R_3 is used to limit the current through the LED for protection purpose. When output is at logic 1, the LED will be lit; otherwise it will be off.

Practically, an op-amp used as a comparator shows several disadvantages as compared with a dedicated single-chip comparator:

(1) Op-amps were originally designed to operate in linear mode with negative feedback, with a lengthy recovery time from saturation. Most op-amps have slew rate limitations for high frequency signals. So, op-amps act as a sloppy comparator with a relatively longer propagation delay.

(2) Because op-amps do not have any internal hysteresis, an external hysteresis network should be built when a slowly-fluctuating input signal is applied.

(3) The quiescent current specification of an op-amp is valid only in linear mode. Some op-amps show a larger quiescent current when used as comparators.

(4) A comparator is inherently designed to produce well-made digital logic, but the op-amp is not. So, compatibility between op-amps and digital logic must be verified.

(5) Several op-amps may be contained within one chip. In this case, cross-channel interaction may exist when used as comparators.

(6) Normally, op-amps have back-to-back diodes between their inputs. Op-amp inputs usually follow each other, and this is fine for amplifiers. However, these diodes may cause unexpected current through inputs when used as comparators.

Fortunately, single-chip comparators are available. Improvements, such as faster switching between output levels, noise immunity to avoid oscillation and stronger output driving ability have been built into them.

5.6.2 Schmitt trigger

In electronics, a Schmitt trigger is a comparator circuit with hysteresis implemented by applying positive feedback to the noninverting input of a comparator or differential amplifier [14].

A Schmitt trigger is an active circuit that converts an analog input signal to a digital output signal. The circuit is called a "trigger" because the output keeps its value until the input changes sufficiently to trigger a change.

For example, in the noninverting configuration, when the input is higher than a chosen high threshold, the output is high. When the input is below a chosen low threshold, the output is low, and when the input is between the two thresholds, the output keeps its previous value.

This dual-threshold action is called "hysteresis" and indicates that the Schmitt trigger possesses some kind of memory and can act as a bistable multivibrator (latch or flip-flop). There exists a tight relation between the Schmitt trigger and the latch; they are interchangeable.

Figure 5.71 shows the comparison of the output of a normal comparator (in the middle) and a Schmitt trigger (at the bottom) with a fluctuating analog input signal (at the top). The comparator has a reference v_{ref}, and the Schmitt trigger has two thresholds, v_{THlow} and v_{THhigh}. The waveform of the comparator changes each time the input signal crosses the reference v_{ref}, leading to many rising and falling edges. On the other hand, the waveform of the Schmitt trigger only changes when the input signal rises across the high threshold v_{THhigh} and falls across the low threshold v_{THlow}. This shows some level of noise tolerance for the input signal

Schmitt triggers are commonly implemented using an operational amplifier or a dedicated single-chip comparator. An open-loop op-amp and comparator may be considered as an analog-digital device with analog inputs and a digital output that extracts the sign of the voltage difference between its two inputs.

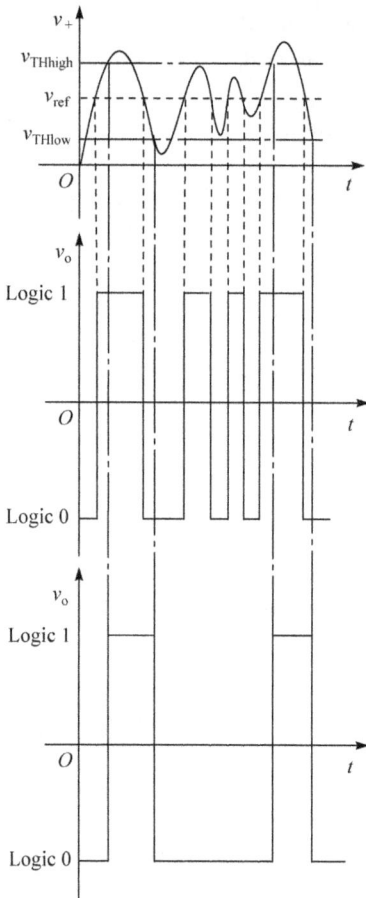

Fig. 5.71: Comparison of output waveforms of a comparator and a Schmitt trigger.

Schmitt triggers are widely used in signal modification applications to remove noise from signals used in digital circuits, particularly mechanical contact bounce in switches. They can also be used in closed loops with negative feedback to implement relaxation oscillators, which can be used as switching power supplies or function generators.

5.6.3 Noninverting Schmitt trigger

Figure 5.72 shows the circuit of the noninverting Schmitt trigger, with the two resistors R_1 and R_f forming a parallel voltage summer. The resistor R_f adds a part of the output voltage v_o to the input voltage v_i. This parallel positive feedback creates the necessary hysteresis manipulated by the ratio between R_1 and R_f. The output of the parallel voltage summer is single ended, with respect to ground, so the inverting input is grounded to make the reference point zero volts [11, 32].

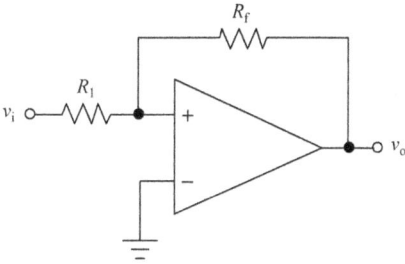

Fig. 5.72: Network of the noninverting Schmitt trigger.

The voltage on the negative input $v_- = 0$ and based on the superposition theorem, by setting either of v_i and v_o to zero, the voltage on the positive input v_+ is obtained as

$$v_+ = \frac{R_f}{R_1 + R_f}v_i + \frac{R_1}{R_1 + R_f}v_o$$

If the input signal v_i is high enough, the output v_o is

$$v_o = V_{oH}$$

Then the positive input signal

$$v_+ = \frac{R_f}{R_1 + R_f}v_i + \frac{R_1}{R_1 + R_f}V_{oH}$$

The positive input voltage v_+ decreases due to the decreasing of input signal v_i. The positive signal v_+ decreases to zero, the same as the negative input signal, $v_+ = v_- = 0$ before the output voltage v_o switches to the low level of V_{oL}. In other words, when the input signal v_i reaches the threshold voltage, we obtain

$$0 = \frac{R_f}{R_1 + R_f}V_{THhigh} + \frac{R_1}{R_1 + R_f}V_{oH}$$

Therefore, the threshold voltage V_{THhigh} for the input signal v_i when the output voltage switches from high to low is

$$V_{THhigh} = -\frac{R_1}{R_f}V_{oH}$$

Note that V_{THhigh} is a negative voltage.

On the other hand, if the input signal v_i is low enough that the output v_o is V_{oL}, that is $v_o = V_{oL}$, the positive input is

$$v_+ = \frac{R_f}{R_1 + R_f}v_i + \frac{R_1}{R_1 + R_f}V_{oL}$$

The input signal v_i increases to $v_i = V_{THlow}$ and the positive input voltage v_+ increases to $v_+ = v_- = 0$ before the output voltage switches to the high level of V_{oH}. Then, we obtain that

$$0 = \frac{R_f}{R_1 + R_f}V_{THlow} + \frac{R_1}{R_1 + R_f}V_{oL}$$

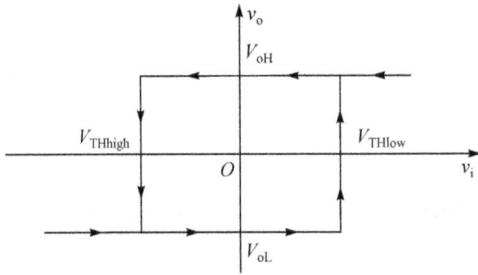

Fig. 5.73: Transfer characteristics of a noninverting Schmitt trigger.

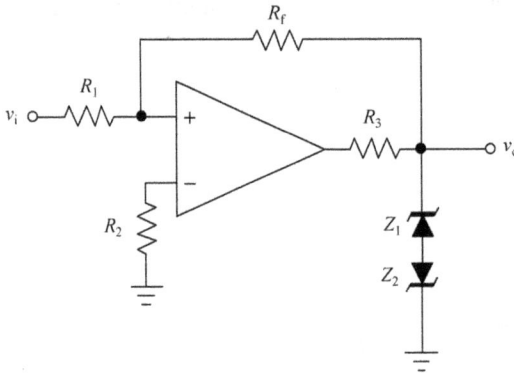

Fig. 5.74: A practical Schmitt trigger circuit with adjustable thresholds.

Therefore, the threshold voltage V_{THlow} for the input signal v_i when the output voltage v_o switches from low to high is

$$v_{THlow} = -\frac{R_1}{R_f} V_{oL}$$

Note that V_{THlow} is a positive voltage because V_{oL} is negative.

Figure 5.73 shows the transfer characteristics of the noninverting Schmitt trigger. The hysteresis is obvious; the switching threshold is different for a rising input from that for a falling input. A practical Schmitt trigger with precise thresholds is shown in Fig. 5.74 [11, 32]. The transfer characteristic has exactly the same shape of Fig. 5.73, and so do the threshold values.

On the other hand, in the circuit shown in Fig. 5.72, the output voltage depends on the power supply, while the output of Fig. 5.74 is defined by the Zener diodes. In this configuration, the output levels can be modified by the appropriate choice of the Zener diode, and these levels are resistant to power supply fluctuations. The resistor R_3 is there to limit the current through the diodes, and the resistor R_2 minimizes the input voltage offset caused by the input leakage currents.

5.6.4 Inverting Schmitt trigger

Figure 5.75 shows the inverting Schmitt trigger [11, 32].

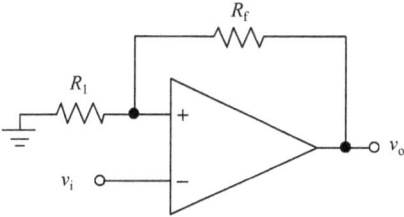

Fig. 5.75: Network of the inverting Schmitt trigger.

The two resistors R_1 and R_f perform the effect of attenuation in the form of a voltage divider. The positive input terminal acts as a simple series voltage summer that adds a part of the output voltage v_o in series to the circuit input voltage v_+. This series positive feedback generates the necessary hysteresis that is manipulated by the ratio between the resistances of R_1 and the whole resistance (R_1 and R_f).

The word "inverting" in the name of the circuit is based on the fact that the output voltage v_o always has an opposite sign to the input voltage v_i when it is out of the hysteresis cycle, that is, when the input voltage v_i is above the high threshold V_{THhigh} or below the low threshold V_{THlow}. However, if within the hysteresis cycle, that is, v_i is between the V_{THhigh} and V_{THlow}, the circuit may be inverting or noninverting, depending on the last state of the changing v_i signal.

The voltage on the negative input is $v_- = v_i$. Based on the superposition theorem, the voltage on the positive input is

$$v_+ = \frac{R_1}{R_1 + R_f} v_o$$

As a comparator, the output voltage v_o is either a high level of V_{oH} or a low level of V_{oL}. If the input signal v_i is low enough that the output $v_o = V_{oH}$, then due to the positive feedback, the positive input signal

$$v_+ = \frac{R_1}{R_1 + R_f} V_{oH}$$

which acts as a reference of the comparator when the output is high.

The input signal v_i increases to a positive input voltage of

$$v_i = \frac{R_1}{R_1 + R_f} V_{oH}$$

before the output voltage v_o switches to the low level of V_{oL}. Therefore, the threshold voltage V_{THhigh} for the input signal v_i when the output voltage switches from high to low is

$$V_{\text{THhigh}} = \frac{R_1}{R_1 + R_f} V_{\text{oH}}$$

On the other hand, if the input signal v_i is high enough that the output v_o is

$$v_o = V_{\text{oL}}$$

then the positive input signal

$$v_+ = \frac{R_1}{R_1 + R_f} V_{\text{oL}}$$

which also acts as a reference of the comparator when the output is low. The input signal v_i decreases to the positive input voltage of

$$v_i = v_+ = \frac{R_1}{R_1 + R_f} V_{\text{oL}}$$

before the output voltage v_o switches to the high level of V_{oH}. Therefore, the threshold voltage V_{THlow} for the input signal v_i when the output voltage switches from low to high is

$$V_{\text{THlow}} = \frac{R_1}{R_1 + R_f} V_{\text{oL}}$$

Figure 5.76 shows the transfer characteristic of the inverting Schmitt trigger [11, 32].

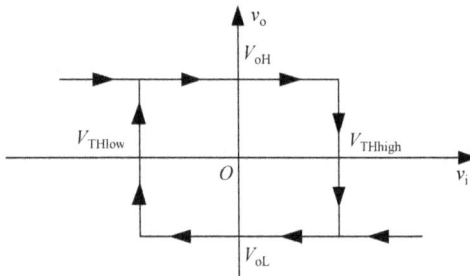

Fig. 5.76: Transfer characteristic of the inverting Schmitt trigger.

Example 5.2
For the inverting Schmitt trigger in Fig. 5.75 the output voltages are ±15 V, $R_1 = 1.5\,\text{k}\Omega$ and $R_f = 3.0\,\text{k}\Omega$.
(1) Sketch the transfer characteristic of the inverting Schmitt trigger.
(2) For the input signal shown in Fig. 5.77, plot the output waveform when output is high at $t = 0$.

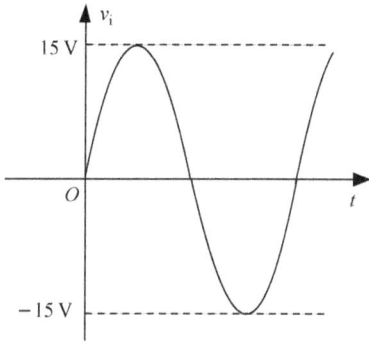

Fig. 5.77: Input signal of Example 5.2.

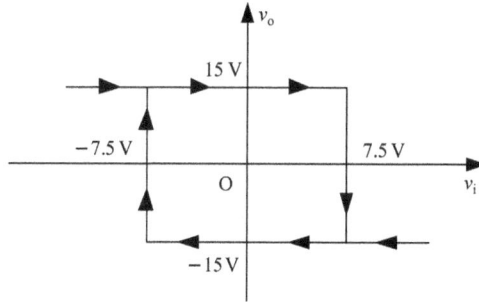

Fig. 5.78: Transfer characteristic of Example 5.2.

Solution

(1) From the above derivations,

$$V_{\text{THhigh}} = \frac{R_1}{R_1 + R_f} V_{\text{oH}}$$

$$= \frac{1.5\,\text{k}\Omega}{1.5\,\text{k}\Omega + 3.0\,\text{k}\Omega} 15\,\text{V}$$

$$= 7.5\,\text{V}$$

$$V_{\text{THlow}} = \frac{R_1}{R_1 + R_f} V_{\text{oL}}$$

$$= \frac{1.5\,\text{k}\Omega}{1.5\,\text{k}\Omega + 3.0\,\text{k}\Omega}(-15\,\text{V})$$

$$= -7.5\,\text{V}$$

The transfer characteristic is shown in Fig. 5.78.

(2) From the obtained threshold voltage V_{THhigh} and V_{THlow}, the output waveform is shown in Fig. 5.79.

5.6.5 Precision rectifier

In electronics, rectification is referred to as the process of separating the positive and negative portions of a waveform and selecting either part to retain. As discussed before, semiconductor diodes can be used in rectification circuits. Furthermore, the precision rectifier, sometimes called the "super diode", is a network implemented by an op-amp for the purpose of setting up a circuit with behavior like an ideal diode. It is useful for signal processing with high precision. The basic circuit is shown in Fig. 5.80, with R_L as the load resistor [33, 34].

When the input voltage v_i is negative, the output of the op-amp v_{o1} is a negative voltage, so the diode D_1 works like an open circuit; no current goes through the load resistor R_L, and the output voltage v_o is zero. Then, when the input is v_i positive, the

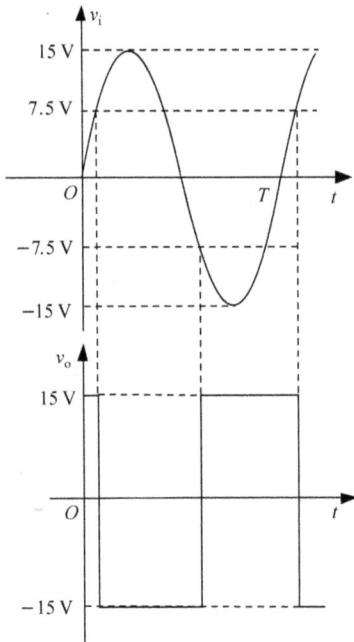

Fig. 5.79: Input and output waveforms of Example 5.2.

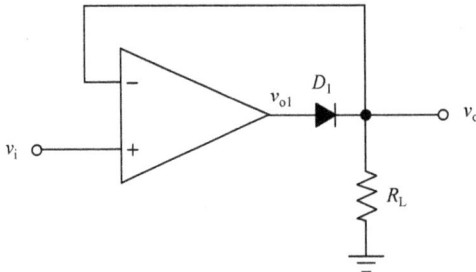

Fig. 5.80: Network of the precision rectifier.

output of the op-amp v_{o1} is also positive, so the diode D_1 works like a short circuit; the current goes through the load resistor R_L. Then the negative feedback works, leading to the output voltage v_o being equal to the input voltage v_i.

The actual threshold V_T of the super diode is very close to zero. It equals the actual threshold V_T of the diode divided by the gain of the op-amp A_v. This makes the super diode much closer to an ideal one.

This basic super diode network is not perfect. When the input is slightly negative, the op-amp runs as an open loop with no feedback signal through the diode. So, the op-amp output saturates with high open-loop gain. Then, when the input changes to positive, the op-amp has to get out of the saturation state before positive amplification works. This takes some time and results in ringing, greatly degrading the frequency response of the circuit. So, the basic super diode network is not widely used. An improved network with two diodes is given in Fig. 5.81 [33, 34].

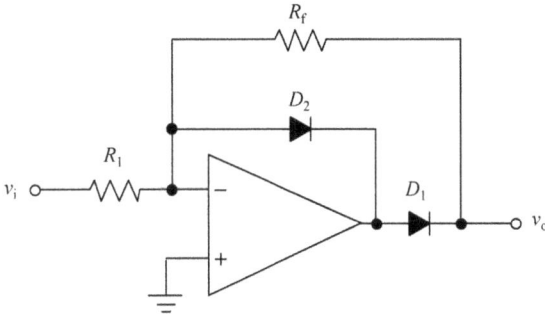

Fig. 5.81: Improved network of the precision rectifier.

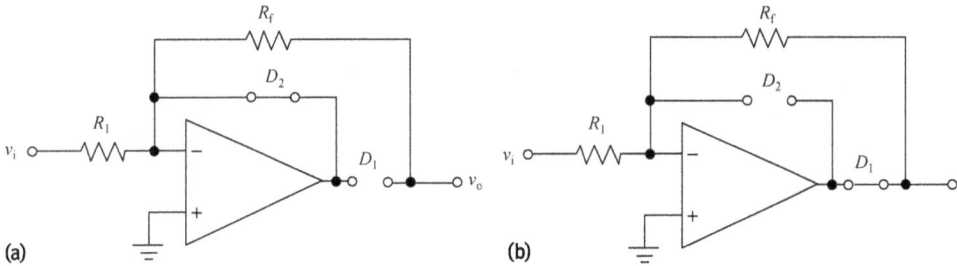

Fig. 5.82: Equivalent circuit of the rectifier, (a) When $v_i > 0$, (b) When $v_i < 0$.

In this configuration, when the input v_i is greater than zero, D_1 is off and D_2 is on, so the output v_o is zero because D_2 sets up a negative feedback and the right-hand side R_f is connected to the virtual ground with no current through it. The equivalent circuit is shown in Fig. 5.82 (a). When the input is negative, D_1 is on and D_2 is off, so the equivalent circuit is the inverting amplifier. Then, the output v_o is the amplification of the input with the ratio of $-R_f/R_1$. The equivalent circuit is shown in Fig. 5.82 (b).

As a result, the output voltage is a true, accurate and inverted reproduction of the negative portions of the input signal v_i. Thus, this circuit operates as a precision half-wave rectifier. If R_f is equal to R_1, as the usual case, the output voltage v_o will have the same amplitude as the input voltage v_i. The waveforms of input v_i and output v_o are given in Fig. 5.83.

On the other hand, if the positive portion of the input signal is needed, simply reverse the two diodes. The result will be a negative waveform, which is the reversed copy of the positive part of the input signal.

This configuration has the benefit that the op-amp never goes into the saturated state, except that its output will have doubled diode threshold voltage drops each time the input signal crosses zero. The configuration to perform full-wave rectification is shown in Fig. 5.84.

Actually, the full-wave rectifier is the combination of a half-wave rectifier and a two-input negative summing amplifier. The waveforms on some important nodes are

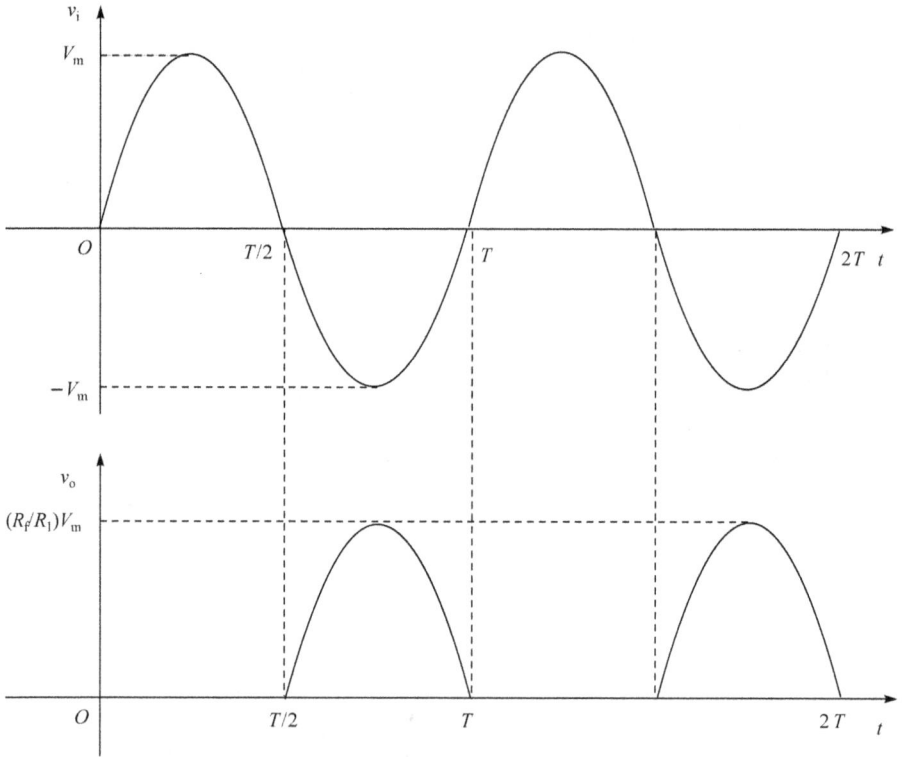

Fig. 5.83: Waveforms of the input and output of the precision rectifier.

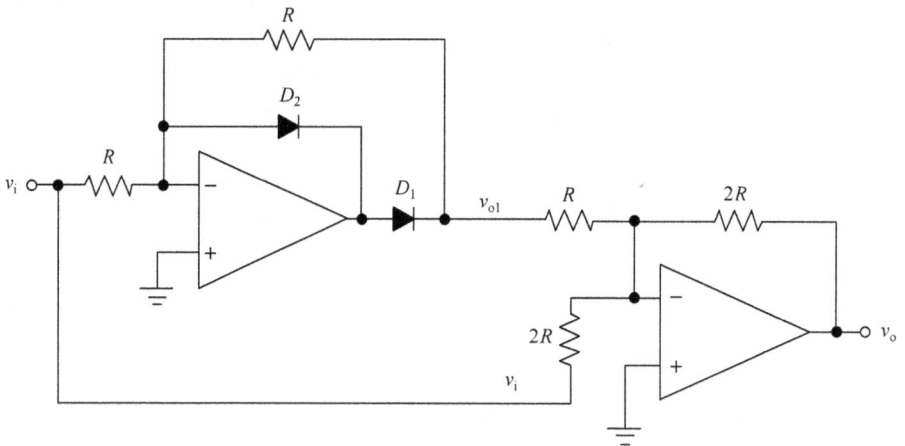

Fig. 5.84: Network of the precision full-wave rectifier.

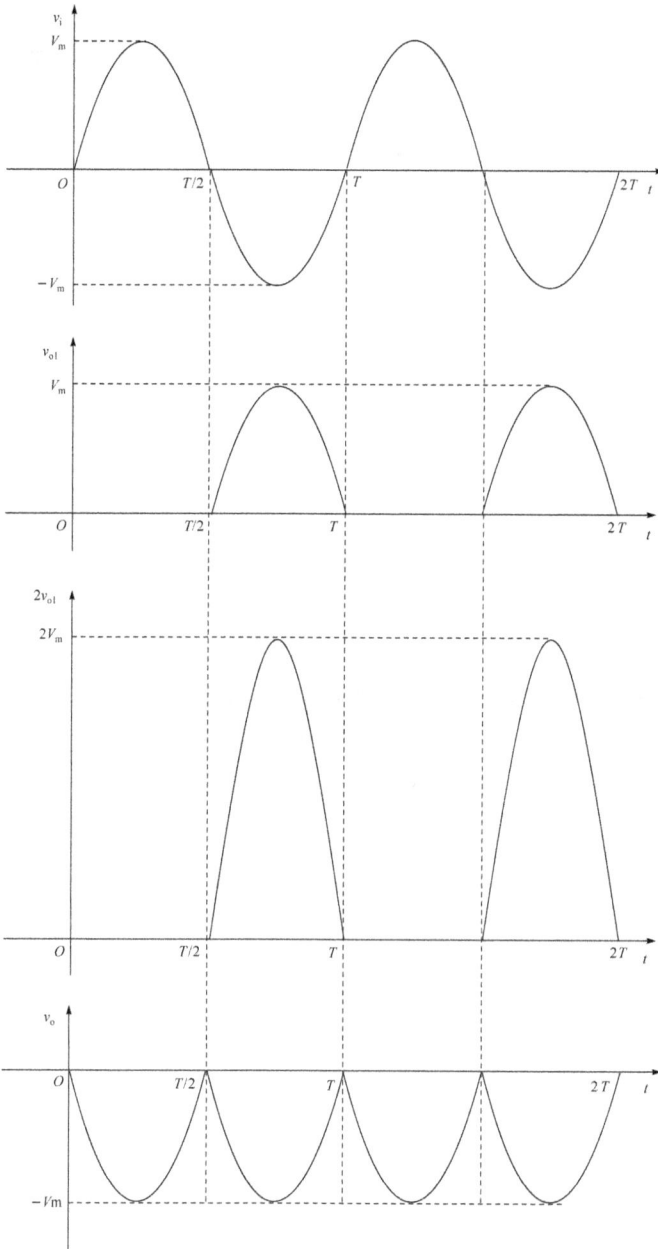

Fig. 5.85: Waveforms of the precision full-wave rectifier.

shown in Fig. 5.85. It can be seen that the output of the half-wave rectifier, v_{o1}, is the reversed copy of the negative part of the input signal v_i. The inverting summing amplifier performs the following operation:

$$v_o = -(2v_{o1} + v_i)$$

So, the result is the combination of the negative copy of the positive half of the input signal v_i and the exact copy of the negative half of the input signal v_i, as the waveform at the bottom of Fig. 5.85. Note that the resistor values are crucial; the resistor values must be of high precision in order to keep the rectification process accurate. Also, the resistance ratio must be maintained as indicated in Fig. 5.85. Moreover, if positive output waveforms are needed, a reversion of the two diodes in the half-wave rectifier will serve the purpose.

5.6.6 Logarithmic amplifier

A logarithmic amplifier is an amplifier of which the output voltage v_o has a natural log relationship with the input voltage v_i. The basic configuration is shown in Fig. 5.86 [35, 36].

Recall the conclusion in Chapter 2 that the general characteristics of a semiconductor diode can be defined by Eq. (2.1), Shockley's equation,

$$I_D = I_S(e^{V_D/nV_T} - 1)$$

where I_S is the reverse saturation current; V_D is the applied forward-bias voltage across the diode, n is assumed as 1, and V_T is the thermal voltage.

In Fig. 5.86 it is also assumed that the op-amp is an ideal one with large A_d and a feedback component D exists. So the golden rules of the op-amp are applicable to analyzing logarithmic amplifiers. Then, the virtual short circuit exits between the two input terminals and virtual ground exists in the negative input terminal, as shown in Fig. 5.87 [35, 36].

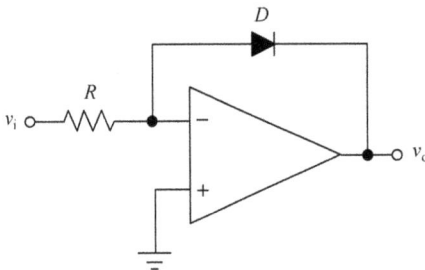

Fig. 5.86: Basic configuration of the logarithmic amplifier.

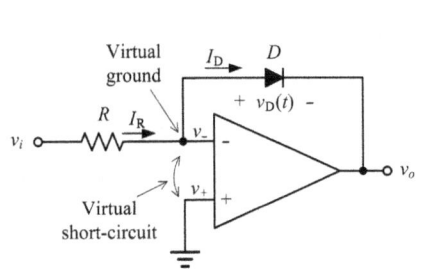

Fig. 5.87: Analysis of the logarithmic amplifier.

Thus, the voltage levels in the two input terminals are the same, that is,

$$V_+ = V_- = 0$$

Then, the currents through R can be obtained,

$$I_R = \frac{v_i}{R}$$

For the diode D, from Shockley's equation, it is obvious that

$$I_D \approx I_S(e^{V_D/V_T})$$

and, in fact, $I_R = I_D$, because no current enters into the op-amp through either input terminal, so,

$$\frac{v_i}{R} = I_S(e^{V_D/V_T})$$

Also, $v_o(t) = -v_D(t)$ with respect to virtual ground, so,

$$\frac{v_i}{R} = I_S(e^{-v_o/V_T})$$

Rearrangement leads to

$$v_o = -V_T \ln \frac{v_i}{I_S R}$$

A necessary condition for successful operation of a log amplifier is that the input voltage v_i should be always positive. This may be accomplished by a rectifier followed by a filter to ensure the polarity of the input signal before applying it to the input. As v_i is positive, v_o is obliged to be negative, because the network is the inverting amplifier.

Another configuration with a transistor can also implement logarithmic operation, as shown in Fig. 5.88 [35, 36]. As v_o is negative and large enough to keep the

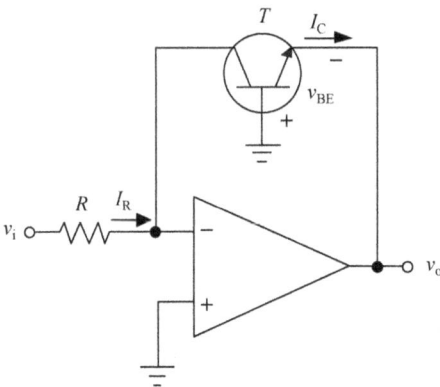

Fig. 5.88: Configuration of logarithmic amplifier with transistor.

emitter-base junction of the BJT forward biased, ON state,

$$I_C = I_S(e^{V_{BE}/V_T} - 1)$$
$$\approx I_S(e^{V_{BE}/V_T})$$
$$= I_S(e^{-V_0/V_T})$$
$$= \frac{V_i}{R}$$

Then,

$$V_0 = -V_T \ln \frac{V_i}{I_S R}$$

Now the output voltage v_o is expressed as the natural log of the input voltage v_i. However, both the saturation current I_S and the thermal voltage V_T are temperature dependent, and thus additional temperature compensating circuits may be needed.

5.6.7 Exponential amplifiers

An exponential amplifier is an amplifier of which the output voltage v_o has an exponential relationship with the input voltage v_i. The basic configuration is shown in Fig. 5.89 [35].

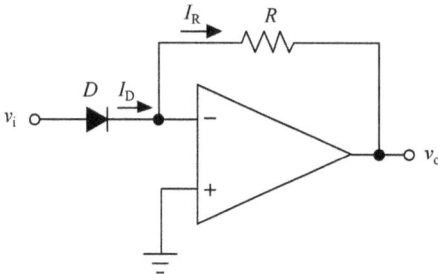

Fig. 5.89: Network of the exponential amplifier.

Based on the properties of diode, it is obvious that

$$I_D = I_S(e^{V_D/V_T} - 1)$$
$$\approx I_S(e^{V_D/V_T})$$
$$= I_S(e^{V_i/V_T})$$

Also, applying the golden rules of the op-amp, we obtain

$$I_D = I_R$$
$$= -\frac{V_o}{R}$$

So,

$$v_o = -RI_D$$
$$= -RI_S(e^{v_i/V_T})$$

Now the output voltage v_o is expressed as the exponential function of the input voltage v_i. Note that the input voltage v_i should be greater than zero.

5.6.8 Oscillators

An electronic oscillator is an electronic network that produces a periodic electronic oscillation, normally a square wave or a sine wave. Actually, oscillators convert from DC power supply to AC signals. They are used in many electronic devices, such as radio and television transmitters that need carrier frequencies for transmission, computer and digital systems that need synchronous clock signals, and electronic beepers and video games that use synthesized sounds or voices [3, 35, 37–39].

By their output frequencies, oscillators can be classified into low-frequency oscillators (LFO), below 20 Hz; audio oscillators in the audio range of about 16–20 kHz, RF oscillators in the radio frequency (RF) of about 100 kHz to 100 GHz. Moreover, from output waveforms, there are sine wave, square wave, triangle wave and sawtooth wave oscillators. From the purposes they serve, there are system clocks, local oscillators, beat frequency oscillators, signal generators and function generators. From the components involved in setting the frequency, there are RC, LC and crystal oscillators. From the available tuning spectrum, there are fixed, adjustable and wide-range oscillators. Also, from the viewpoint of oscillation principles, oscillators may be classified into linear or harmonic oscillators and nonlinear or relaxation oscillators.

The linear oscillator is the most common form. It is an electronic amplifier, such as an op-amp amplifier, with its output extracted and sent back to the input terminal and added with the input signal, thus called positive feedback, as shown in Fig. 5.90.

Once the power supply to the amplifier is turned on, electronic noise in the input terminal is amplified, and this amplified noise travels around the feedback loop and comes back into the input terminal. Then, it is amplified and sent back again until it very quickly converges. The output voltage can change slowly as it approaches the power supply rail, and then gradually switches its direction. This behavior produces a sine wave of single frequency.

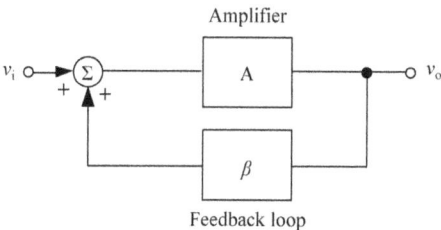

Fig. 5.90: Linear oscillator with positive feedback.

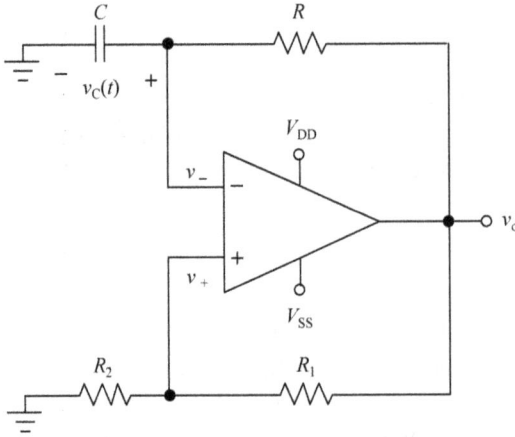

Fig. 5.91: Comparator-based relaxation oscillator.

On the other hand, the output voltage may hover for a time, probably as a process of capacitor discharging, and then rapidly turn to the other power supply rail. Thus, the output takes the form of a square or rectangular wave. This type of oscillator is called a relaxation oscillator. The typical circuit of a relaxation oscillator is shown in Fig. 5.91 [3, 35].

First, it is assumed that output voltage v_o

$$v_o = V_{DD}$$

and the capacitor C has the initial voltage

$$v_C(0) = 0$$

So, in this condition, the capacitor C is being charged to the high output voltage V_{DD}. The positive input terminal of the op-amp is

$$v_+ = \frac{R_2}{R_1 + R_2} V_{DD}$$

This voltage is just the reference voltage.

When the capacitor C has been charged to the reference voltage, that is,

$$v_C = v_- = v_+ = \frac{R_2}{R_1 + R_2} V_{DD}$$

the output voltage v_o will switch from high to low. Then when v_o is low, the capacitor C is discharged to the low output voltage V_{SS}, which is a negative voltage. At this time, the reference voltage is

$$v_+ = \frac{R_2}{R_1 + R_2} V_{SS}$$

When the capacitor C discharges to this reference voltage, the output voltage v_o switches from low to high. Thus, a period completes and the next period begins.

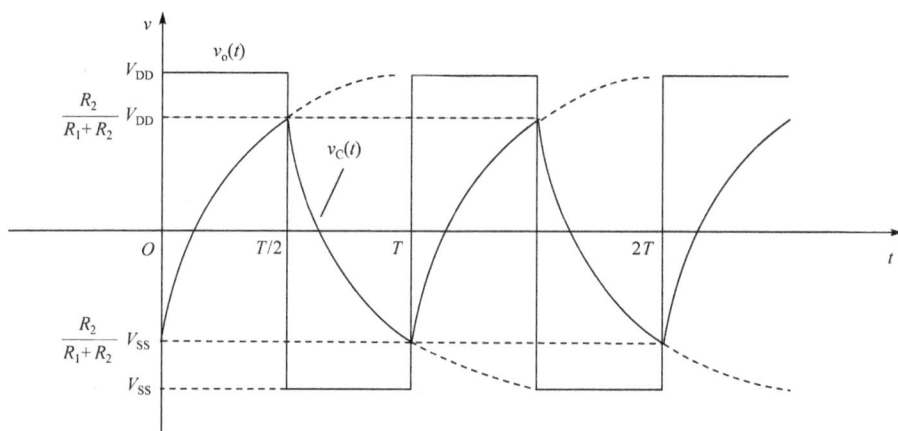

Fig. 5.92: Waveform of the relaxation oscillator.

The output voltage v_o is either low, V_{SS} or high, V_{DD}, a rectangular waveform. The capacitor voltage v_C is between $V_{SS} \cdot R_2/(R_1 + R_2)$ and $V_{DD} \cdot R_2/(R_1 + R_2)V_{DD}$, an approximation of a triangular waveform, as shown in Fig. 5.92.

The frequency of the rectangular output v_o can be obtained by

$$f_o = \frac{1}{2RC \ln\left(1 + \frac{2R_2}{R_1}\right)}$$

So, by properly selecting the values of R_1, R_2, R and C, the output frequency and the duty cycle can be manipulated.

5.7 Chapter summary

Concepts and conclusions
(1) A basic op-amp has two input terminals and one output terminal. The input terminal indicated by a plus (+) sign results in an output with the same polarity as that of the input, while one with a minus (–) sign in the opposite polarity of the input signal. See Section 5.2.
(2) When one input terminal is fed with input signal, while the other is connected to ground, the op-amp is in single-ended input mode. See Section 5.2.
(3) When two input terminals of the op-amp are used for input signals, the op-amp is in double-ended input mode. See Section 5.2.
(4) The op-amp can also have two output terminals with opposite outputs, resulting in a double-ended output mode. See Section 5.2.
(5) When the same input signals are applied to both inputs, the op-amp is in the common-mode operation mode. See Section 5.2.

(6) Signals that are common at both inputs are only slightly amplified, while those that are opposite between the two inputs are highly amplified. This operating feature is referred to as common-mode rejection. See Section 5.2.

(7) In Section 5.3, the common-mode rejection ratio (CMRR) can be calculated as follows:

$$CMMR = \frac{A_d}{A_c}$$

Also, in logarithmic form

$$CMMR(dB) = 20\log_{10}\left(\frac{A_d}{A_c}\right)$$

(8) The relationship between output voltage v_o and differential input v_d is described by the op-amp transfer characteristic, and an equivalent circuit can be developed. See Section 5.4.1.

(9) In Section 5.4.2, the basic op-amp network is a constant-gain amplifier with the voltage gain

$$\frac{v_o}{v_i} = -\frac{R_f}{R_1} \qquad\qquad \text{Eq. (5.4)}$$

(10) The concepts of virtual ground and virtual short circuit are the golden rules of op-amps. See Section 5.4.3.

(11) In Section 5.4.4, basic op-amp circuits include the following:
A unity gain circuit with voltage gain:

$$\frac{v_o}{v_i} = -1$$

A constant gain circuit with voltage gain:

$$\frac{v_o}{v_i} = -k$$

An inverting amplifier with voltage gain:

$$\frac{v_o}{v_i} = -\frac{R_f}{R_1}$$

A non-inverting amplifier with voltage gain:

$$\frac{v_o}{v_i} = \left(1 + \frac{R_f}{R_1}\right)$$

Unity follower with voltage gain:

$$\frac{v_o}{v_i} = 1$$

(12) In Section 5.5, typical linear op-amp application circuits include the following:
Voltage summing with the output voltage:

$$v_o = -\frac{R_f}{R_1}v_{i1} - \frac{R_f}{R_2}v_{i2} - \frac{R_f}{R_3}v_{i3}$$

Voltage subtraction with the output voltage:

$$v_o = \frac{R_3}{R_2 + R_3} \cdot \frac{R_1 + R_f}{R_1} v_{i2} - \frac{R_f}{R_1} v_{i1}$$

An integrator with the output voltage:

$$v_o(t) = -\frac{1}{RC} \int v_i(t)dt$$

A differentiator with the output voltage:

$$v_o = -RC\frac{dv_i(t)}{dt}$$

An instrumentation amplifier with the output voltage:

$$v_o = \frac{R_3}{R_2}\left(1 + \frac{2R_1}{R_G}\right)(v_{i2} - v_{i1})$$

(13) Active filters, such as low-pass filters, high-pass filters and bandpass filters can be implemented by op-amps. See Section 5.5.6.

(14) In Section 5.6, some typical nonlinear op-amp application circuits include the following:

- A comparator, implemented by a high-gain differential amplifier in the open-loop with bipolar supply, compares two input voltages and outputs a digital signal indicating which is larger. See Section 5.6.1.
- A Schmitt trigger is a comparator circuit with hysteresis implemented by applying positive feedback to the noninverting input of a comparator or differential amplifier. It can also be classified into inverting and noninverting Schmitt triggers. See Section 5.6.2, 5.6.3 and 5.6.4.
- A rectification circuit, implemented by an op-amp, can separate the positive and negative portions of a waveform and select to retain either part. See Section 5.6.5.
- A logarithmic amplifier is an amplifier of which the output voltage v_o has a natural log relationship with the input voltage v_i. See Section 5.6.6.
- An exponential amplifier implements that the output voltage v_o has an exponential relationship with the input voltage v_i. See Section 5.6.7.
- An oscillator is an electronic network that produces a periodic electronic oscillation. If the output takes the form of a square or rectangular wave, it is called the relaxation oscillator. See Section 5.6.8.

5.8 Questions

Q5.1:   The amplification ratio of an op-amp is very large. What is the largest the output of an op-amp can be?

Q5.2: An op-amp has positive and negative input terminals. Recall the three configurations of BJTs, each of which has one input terminal. Are they positive or negative? Why?

Q5.3: What is the single-ended input mode? What is the double-ended input mode with one input? What is the main difference of the input signals between the two modes?

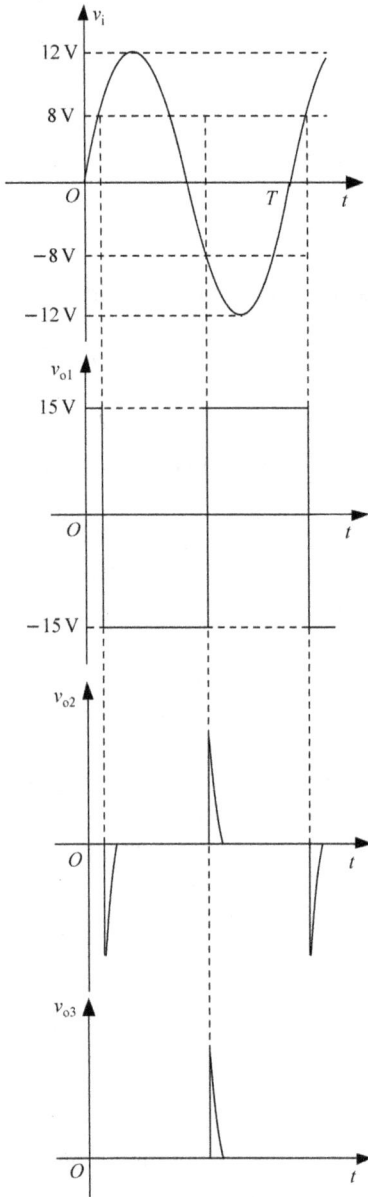

Fig. 5.93: Circuit for Q5.12.

Q5.4: The double-ended output mode can have one or two outputs. What is the difference?

Q5.5: What is the practical significance of the op-amp common-mode rejection?

Q5.6: What is the shortcoming of the op-amp when used in open loops with a very narrow linear region?

Q5.7: What measure can be used to extend the linear region?

Q5.8: What are the golden rules of op-amps? In what conditions?

Q5.9: What is the counterpart of the unity follower circuit of op-amps for BJTs and JFETs?

Q5.10: In the transfer characteristics of op-amps, what is different for linear and nonlinear applications?

Q5.11: What is different in op-amp output voltages for linear and nonlinear applications?

Q5.12: Design a network that can implement the output waveforms in Fig. 5.93.

Bibliography

[1] Thompson MT. Intuitive analog circuit design. 1st edn. Oxford: Newnes; 2006.
[2] Luecke J. Analog and digital circuits for electronic control system applications. 1st edn. Amsterdam: Elsevier; 2005.
[3] Lal Kishore K. Electronic devices and circuits. 1st edn. Hyderabad: BS Publications; 2008.
[4] Lal Kishore K. Electronic circuit analysis. 2nd edn. Hyderabad: BS Publications; 2008.
[5] https://en.wikipedia.org/wiki/Semiconductor_device, last access June 16, 2018.
[6] https://en.wikipedia.org/wiki/Cathode_ray_tube, last access June 16, 2018.
[7] https://en.wikipedia.org/wiki/Vacuum_diode#Diodes, last access June 16, 2018.
[8] https://en.wikipedia.org/wiki/Resistivity, last access June 16, 2018.
[9] Bird J. Electrical circuit theory and technology. Revised 2nd edn. Oxford: Newnes; 2003.
[10] Levinshtein M, Simin G. Transistor: from crystal to integrated circuit. 1st edn. Singapore: World Scientific; 1998.
[11] Boylestad R, Nashelsky L. Electronic devices and circuit theory, 9th edn. London: Pearson Education; 2006.
[12] Staras S. Semiconductor electronic devices study book. 1st edn. Vilnius, Lithuania: Vilnius Gediminas Technical University; 2010.
[13] Hamilton S. An analog electronics companion. 1st edn. Cambridge: Cambridge University Press; 2003.
[14] https://en.wikipedia.org/wiki/Schottky_diode, last access June 16, 2018.
[15] https://en.wikipedia.org/wiki/Schottky_barrier, last access June 16, 2018.
[16] Williams T. The circuit designer's companion. 2nd edn. Oxford, UK: Newnes; 2005.
[17] https://en.wikipedia.org/wiki/Light-emitting_diode, last access June 16, 2018.
[18] https://en.wikipedia.org/wiki/Seven-segment_display, last access June 16, 2018.
[19] https://en.wikipedia.org/wiki/OLED, last access June 16, 2018.
[20] https://en.wikipedia.org/wiki/Vacuum_tube, last access June 16, 2018.
[21] https://en.wikipedia.org/wiki/Operational_amplifier, last access June 16, 2018.
[22] https://en.wikipedia.org/wiki/Integrator, last access June 16, 2018.
[23] https://en.wikipedia.org/wiki/Differentiator, last access June 16, 2018.
[24] https://en.wikipedia.org/wiki/Instrumentation_amplifier, last access June 16, 2018.
[25] http://www.play-hookey.com/analog/feedback_circuits/instrument_amp.html, last access June 16, 2018.
[26] http://www.ibiblio.org/kuphaldt/electricCircuits/Semi/SEMI_8.html#xtocid36229, last access June 16, 2018.
[27] https://en.wikipedia.org/wiki/Active_filter, last access June 16, 2018.
[28] https://en.wikipedia.org/wiki/Band-stop_filter, last access June 16, 2018.
[29] https://en.wikipedia.org/wiki/Low-pass_filter, last access June 16, 2018.
[30] https://en.wikipedia.org/wiki/High-pass_filter, last access June 16, 2018.
[31] https://en.wikipedia.org/wiki/Band-pass_filter, last access June 16, 2018.
[32] https://en.wikipedia.org/wiki/Schmitt_trigger, last access June 16, 2018.
[33] https://en.wikipedia.org/wiki/Precision_rectifier, last access June 16, 2018.
[34] http://www.play-hookey.com/analog/feedback_circuits/half-wave_rectifier.html, last access June 16, 2018.
[35] https://en.wikipedia.org/wiki/Operational_amplifier_applications, last access June 16, 2018.
[36] https://en.wikipedia.org/wiki/Log_amplifier, last access June 16, 2018.
[37] https://en.wikipedia.org/wiki/Electronic_oscillator, last access June 16, 2018.
[38] https://en.wikipedia.org/wiki/Relaxation_oscillator, last access June 16, 2018.
[39] http://www.play-hookey.com/oscillators/intro/classify_oscillators.html, last access June 16, 2018.

https://doi.org/10.1515/9783110593860-006

Index

AC equivalent circuit 76
AC parameters 77
AC response 74
active filter 235
Alpha (α) 48
analog signals 1
AND logic 29
approximation method 68

bandpass filter 235
band-stop filter 235
base (B) 42
Beta (β) 50
bipolar junction transistor (BJT) 42

channel resistance 112
circuit model 76
collector (C) 42
Common-base configuration 45
Common-collector configuration 51
Common-emitter configuration 48
Common-mode operation 211
Common-mode rejection 211
common-mode rejection ratio (CMRR) 211
comparators 244
conductors 6
constant gain circuit of op-amp 222
current-controlled device 107

depletion region 13
depletion region of characteristics 122
differentiator of op-amp 232
digital signal 3
double-ended input mode 208
double-ended output mode 209
drain (D) 110
drain characteristics 126

electron 8
emitter (E) 42
emitter bias circuit 62
emitter-follower configuration 94
enhancement region of characteristics 122
Equivalent method 66
exponential amplifiers 260
extrinsic semiconductors 8

FET AC equivalent circuit 167
field-effect transistor 107
fixed-bias circuit 55
fixed-bias configuration for JFETs 129
forward-bias condition 15
full-wave rectifier 34

golden rules of op-amps 220
graphical approach for JFET 117

half-wave rectifier 31
high-pass filter 235
hole 9

Input characteristics 46
input current 77
input impedance 77
input voltage 77
instrumentation amplifier of op-amp 233
insulators 6
integrator of op-amp 229
internal resistance (r_d) 186
intrinsic semiconductors 7
inverting amplifier of op-amp 223
inverting Schmitt trigger 251

junction field-effect transistor 108

light-emitting diode 18
logarithmic amplifier 258
logic circuit 26
low-pass filter 235

majority carrier 10
minority carrier 10
MOSFET 108

n-channel depletion type MOSFET 108
n-channel enhancement type MOSFET 108
n-channel JFET 108
no bias condition 14
noninverting amplifier of op-amp 223
noninverting Schmitt trigger 248
npn transistor 42
n-type material 8

https://doi.org/10.1515/9783110593860-007

operation point 52
operational amplifier (op-amp) 206
OR logic 26
organic light-emitting diode 19
oscillators 261
output characteristics 47
output current 77
output impedance 77
output voltage 77

passive filter 235
p-channel enhancement type MOSFET 108
p-channel JFET 108
pinch-off 112
pinch-off voltage 112
p-n junction 13
pnp transistor 42
precision rectifier 253
p-type material 9

Q-point 52
quiescent point 52

r_e model 79
resistivity 6
reverse-bias condition 14

Schmitt trigger 247
Schottky diode 17

semiconductor diode 14
semiconductors 6
Shockley's equation for diodes 17
Shockley's equation for JFETs 115
shorthand method for JFET 116
silicon dioxide layer 119
single-ended input mode 207
source (S) 110

threshold voltage 125
transconductance (g_m) 165
transfer characteristic for op-amp 216
transfer characteristics for JFET 115
transfer characteristics of a comparator 245
transistor 42

unipolar device 108
unity follower of op-amp 225
unity gain circuit of op-amp 222

vacuum tube 41
voltage subtraction of op-amp 227
voltage summing of op-amp 226
voltage-controlled device 107
voltage-divider bias circuit 66

Zener diodes 21
Zener region 21

* 9 7 8 3 1 1 0 5 9 5 4 0 6 *